Learn

Eureka Math®
Grade 5
Module 6

Published by Great Minds®.

Copyright © 2018 Great Minds®.

Printed in the U.S.A.
This book may be purchased from the publisher at eureka-math.org.
10 9 8 7 6 5 4 3

ISBN 978-1-64054-074-3

G5-M6-L-05.2018

Learn ◆ Practice ◆ Succeed

Eureka Math® student materials for *A Story of Units®* (K–5) are available in the *Learn, Practice, Succeed* trio. This series supports differentiation and remediation while keeping student materials organized and accessible. Educators will find that the *Learn, Practice,* and *Succeed* series also offers coherent—and therefore, more effective—resources for Response to Intervention (RTI), extra practice, and summer learning.

Learn

Eureka Math Learn serves as a student's in-class companion where they show their thinking, share what they know, and watch their knowledge build every day. *Learn* assembles the daily classwork—Application Problems, Exit Tickets, Problem Sets, templates—in an easily stored and navigated volume.

Practice

Each *Eureka Math* lesson begins with a series of energetic, joyous fluency activities, including those found in *Eureka Math Practice.* Students who are fluent in their math facts can master more material more deeply. With *Practice,* students build competence in newly acquired skills and reinforce previous learning in preparation for the next lesson.

Together, *Learn* and *Practice* provide all the print materials students will use for their core math instruction.

Succeed

Eureka Math Succeed enables students to work individually toward mastery. These additional problem sets align lesson by lesson with classroom instruction, making them ideal for use as homework or extra practice. Each problem set is accompanied by a Homework Helper, a set of worked examples that illustrate how to solve similar problems.

Teachers and tutors can use *Succeed* books from prior grade levels as curriculum-consistent tools for filling gaps in foundational knowledge. Students will thrive and progress more quickly as familiar models facilitate connections to their current grade-level content.

Students, families, and educators:

Thank you for being part of the *Eureka Math®* community, where we celebrate the joy, wonder, and thrill of mathematics.

In the *Eureka Math* classroom, new learning is activated through rich experiences and dialogue. The *Learn* book puts in each student's hands the prompts and problem sequences they need to express and consolidate their learning in class.

What is in the Learn book?

Application Problems: Problem solving in a real-world context is a daily part of *Eureka Math*. Students build confidence and perseverance as they apply their knowledge in new and varied situations. The curriculum encourages students to use the RDW process—Read the problem, Draw to make sense of the problem, and Write an equation and a solution. Teachers facilitate as students share their work and explain their solution strategies to one another.

Problem Sets: A carefully sequenced Problem Set provides an in-class opportunity for independent work, with multiple entry points for differentiation. Teachers can use the Preparation and Customization process to select "Must Do" problems for each student. Some students will complete more problems than others; what is important is that all students have a 10-minute period to immediately exercise what they've learned, with light support from their teacher.

Students bring the Problem Set with them to the culminating point of each lesson: the Student Debrief. Here, students reflect with their peers and their teacher, articulating and consolidating what they wondered, noticed, and learned that day.

Exit Tickets: Students show their teacher what they know through their work on the daily Exit Ticket. This check for understanding provides the teacher with valuable real-time evidence of the efficacy of that day's instruction, giving critical insight into where to focus next.

Templates: From time to time, the Application Problem, Problem Set, or other classroom activity requires that students have their own copy of a picture, reusable model, or data set. Each of these templates is provided with the first lesson that requires it.

Where can I learn more about Eureka Math resources?

The Great Minds® team is committed to supporting students, families, and educators with an ever-growing library of resources, available at eureka-math.org. The website also offers inspiring stories of success in the *Eureka Math* community. Share your insights and accomplishments with fellow users by becoming a *Eureka Math* Champion.

Best wishes for a year filled with aha moments!

Jill Diniz

Jill Diniz
Director of Mathematics
Great Minds

The Read–Draw–Write Process

The *Eureka Math* curriculum supports students as they problem-solve by using a simple, repeatable process introduced by the teacher. The Read–Draw–Write (RDW) process calls for students to

1. Read the problem.

2. Draw and label.

3. Write an equation.

4. Write a word sentence (statement).

Educators are encouraged to scaffold the process by interjecting questions such as

- What do you see?

- Can you draw something?

- What conclusions can you make from your drawing?

The more students participate in reasoning through problems with this systematic, open approach, the more they internalize the thought process and apply it instinctively for years to come.

Contents

Module 6: Problem Solving with the Coordinate Plane

A landscaper is planting some marigolds in a row. The row is 2 yards long. The flowers must be spaced $\frac{1}{3}$ yard apart so that they will have proper room to grow. The landscaper plants the first flower at 0. Place points on the number line to show where the landscaper should place the other flowers. How many marigolds will fit in this row?

0   1 yd   2 yd

_____

_____

_____

_____

Read   Draw   Write

EUREKA
MATH

Name _____   Date _____

1. Each shape was placed at a point on the number line s. Give the coordinate of each point below.

    a.  _____   b. ★ _____

    c. ⬤ _____   d. ⬛ _____

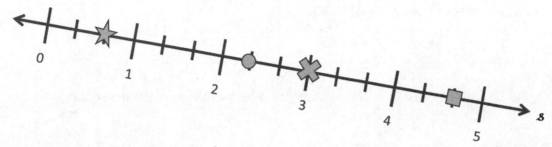

2. Plot the points on the number lines.

    a.

    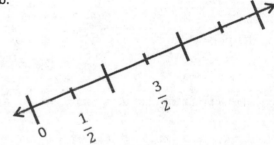

    Plot A so that its distance from the origin is 2.

    b.

    Plot R so that its distance from the origin is $\frac{5}{2}$.

c.

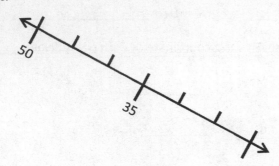

Plot L so that its distance from the origin is 20.

d.

Plot a point T so that its distance from the origin is $\frac{2}{3}$ more than that of S.

3. Number line g is labeled from 0 to 6. Use number line g below to answer the questions.

a. Plot point A at $\frac{3}{4}$.

b. Label a point that lies at $4\frac{1}{2}$ as B.

c. Label a point, C, whose distance from zero is 5 more than that of A.

The coordinate of C is _____.

d. Plot a point, D, whose distance from zero is $1\frac{1}{4}$ less than that of B.

The coordinate of D is _____.

e. The distance of E from zero is $1\frac{3}{4}$ more than that of D. Plot point E.

f. What is the coordinate of the point that lies halfway between A and D? _____
Label this point F.

Lesson 1:   Construct a coordinate system on a line.

EUREKA MATH

4. Mrs. Fan asked her fifth-grade class to create a number line. Lenox created the number line below:

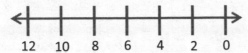

Parks said Lenox's number line is wrong because numbers should always increase from left to right. Who is correct? Explain your thinking.

5. A pirate marked the palm tree on his treasure map and buried his treasure 30 feet away. Do you think he will be able to easily find his treasure when he returns? Why or why not? What might he do to make it easier to find?

Look for the treasure 30 feet from this tree!

Name _____   Date _____

Use number line ℓ to answer the questions.

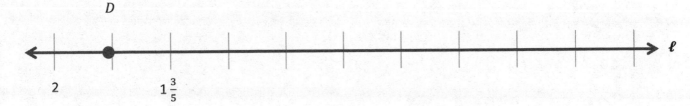

a. Plot point C so that its distance from the origin is 1.

b. Plot point E $\frac{4}{5}$ closer to the origin than C. What is its coordinate? _____

c. Plot a point at the midpoint of C and E. Label it H.

The picture shows an intersection in Stony Brook Village.

a. The town wants to construct two new roads, Elm Street and King Street. Elm Street will intersect Lower Sheep Pasture Road, run parallel to Main Street, and be perpendicular to Stony Brook Road. Sketch Elm Street.

b. King Street will be perpendicular to Main Street and begin at the intersection of Upper Sheep Pasture Road and East Main Street. Sketch King Street.

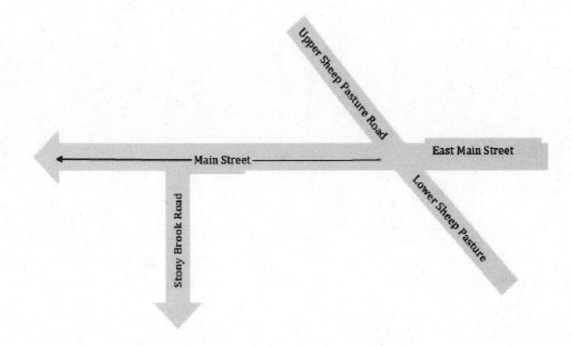

Read   **Draw**   **Write**

Name _____　　　Date _____

1.

a. Use a set square to draw a line perpendicular to the x-axes through points P, Q, and R. Label the new line as the y-axis.

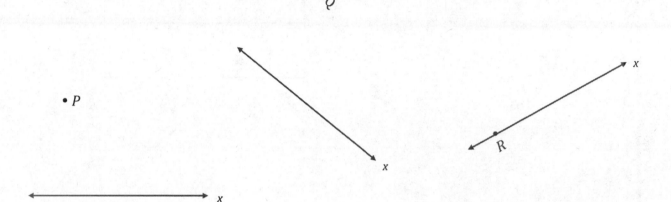

a. Choose one of the sets of perpendicular lines above, and create a coordinate plane. Mark 7 units on each axis, and label them as whole numbers.

2. Use the coordinate plane to answer the following.

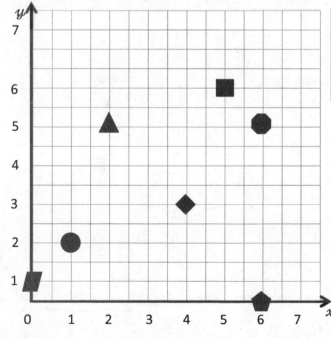

a. Name the shape at each location.

x-coordinate	y-coordinate	Shape
2	5	
1	2	
5	6	
6	5	

b. Which shape is 2 units from the y-axis?

c. Which shape has an x-coordinate of 0?

d. Which shape is 4 units from the y-axis and 3 units from the x-axis?

3. Use the coordinate plane to answer the following.

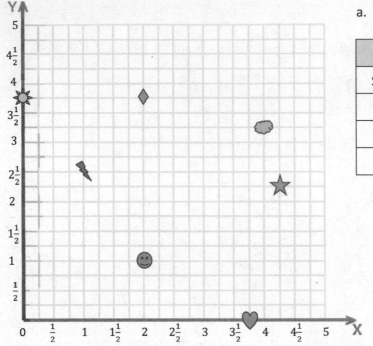

a. Fill in the blanks.

Shape	x-coordinate	y-coordinate
Smiley Face	2	1
Diamond	2	3 3/4
Sun		
Heart		

b. Name the shape whose x-coordinate is $\frac{1}{2}$ more than the value of the heart's x-coordinate.

c. Plot a triangle at (3, 4).

d. Plot a square at ($4\frac{3}{4}$, 5).

e. Plot an X at ($\frac{1}{2}$, $\frac{3}{4}$).

4. The pirate's treasure is buried at the ✖ on the map. How could a coordinate plane make describing its location easier?

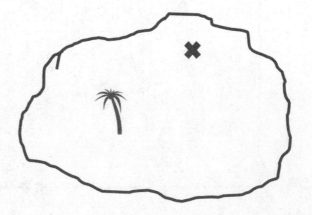

EUREKA
MATH

Name _____   Date _____

1. Name the coordinates of the shapes below.

Shape	x-coordinate	y-coordinate
Sun	1	4,5
Arrow	1.5	2
Heart	4	4.5

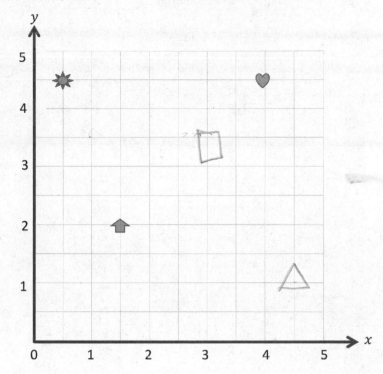

2. Plot a square at $(3, 3\frac{1}{2})$.

3. Plot a triangle at $(4\frac{1}{2}, 1)$.

is it kayle?

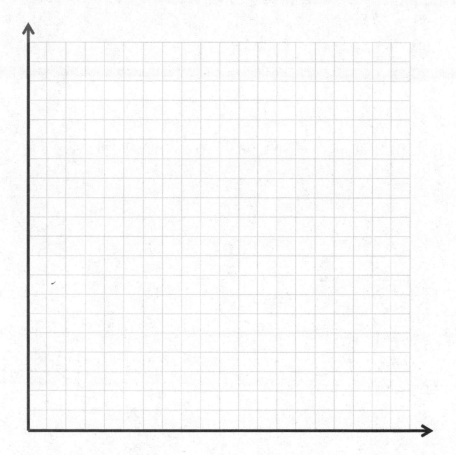

coordinate plane

The captain of a ship has a chart to help him navigate through the islands. He must follow points that show the deepest part of the channel. List the coordinates the captain needs to follow in the order he will encounter them.

1. (_____, _____) 2. (_____, _____)

3. (_____, _____) 4. (_____, _____)

5. (_____, _____) 6. (_____, _____)

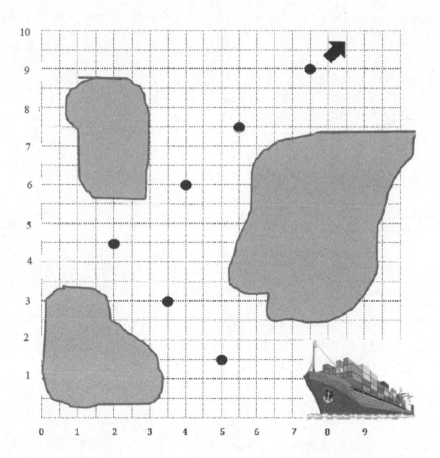

Read   Draw   Write

Lesson 3:   Name points using coordinate pairs, and use the coordinate pairs to plot points.

17

Name _____   Date _____

1. Use the grid below to complete the following tasks.

    a. Construct an x-axis that passes through points A and B.

    b. Construct a perpendicular y-axis that passes through points C and F.

    c. Label the origin as 0.

    d. The x-coordinate of B is $5\frac{2}{3}$. Label the whole numbers along the x-axis.

    e. The y-coordinate of C is $5\frac{1}{3}$. Label the whole numbers along the y-axis.

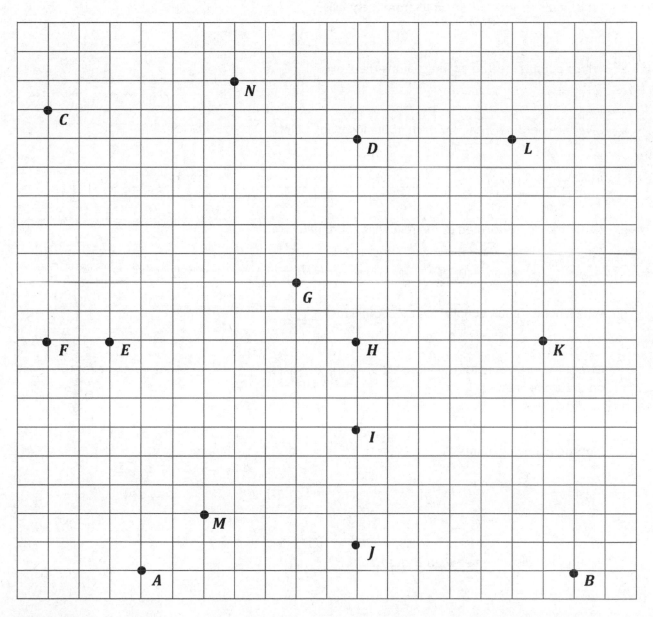

Lesson 3:   Name points using coordinate pairs, and use the coordinate pairs to plot points.

19

© 2018 Great Minds®. eureka-math.org

2. For all of the following problems, consider the points A through N on the previous page.

   a. Identify all of the points that have an x-coordinate of $3\frac{1}{3}$.

   b. Identify all of the points that have a y-coordinate of $2\frac{2}{3}$.

   c. Which point is $3\frac{1}{3}$ units above the x-axis *and* $2\frac{2}{3}$ units to the right of the y-axis? Name the point, and give its coordinate pair.

   d. Which point is located $5\frac{1}{3}$ units from the y-axis?

   e. Which point is located $1\frac{2}{3}$ units along the x-axis?

   f. Give the coordinate pair for each of the following points.

      K: _____ I: _____ B: _____ C: _____

   g. Name the points located at the following coordinates.

      $(1\frac{2}{3}, \frac{2}{3})$ _____ $(0, 2\frac{2}{3})$ _____ $(1, 0)$ _____ $(2, 5\frac{2}{3})$ _____

   h. Which point has an equal x- and y-coordinate? _____

   i. Give the coordinates for the intersection of the two axes. (____, ____) Another name for this point on the plane is the _____.

   j. Plot the following points.

      $P: (4\frac{1}{3}, 4)$ $Q: (\frac{1}{3}, 6)$ $R: (4\frac{2}{3}, 1)$ $S: (0, 1\frac{2}{3})$

   k. What is the distance between E and H, or EH?

Lesson 3:   Name points using coordinate pairs, and use the coordinate pairs to plot points.

© 2018 Great Minds®. eureka-math.org

EUREKA
MATH®

l.   What is the length of HD?

m.   Would the length of ED be greater or less than $EH + HD$?

n.   Jack was absent when the teacher explained how to describe the location of a point on the coordinate plane. Explain it to him using point J.

Lesson 3:   Name points using coordinate pairs, and use the coordinate pairs to
              plot points.

© 2018 Great Minds®. eureka-math.org

21

Name _____   Date _____

Use a ruler on the grid below to construct the axes for a coordinate plane. The x-axis should intersect points L and M. Construct the y-axis so that it contains points K and L. Label each axis.

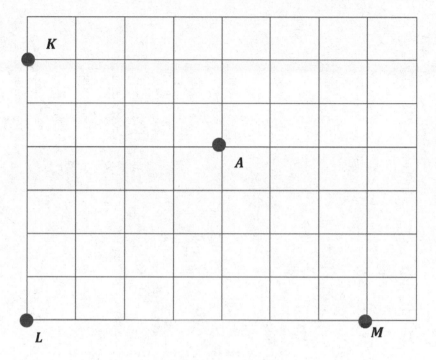

a.   Place a hash mark on each grid line on the x- and y-axis.

b.   Label each hash mark so that A is located at (1, 1).

c.   Plot the following points:

Point	x-coordinate	y-coordinate
B	$\frac{1}{4}$	0
C	$1\frac{1}{4}$	$\frac{3}{4}$

Lesson 3:   Name points using coordinate pairs, and use the coordinate pairs to plot points.

23

© 2018 Great Minds®. eureka-math.org

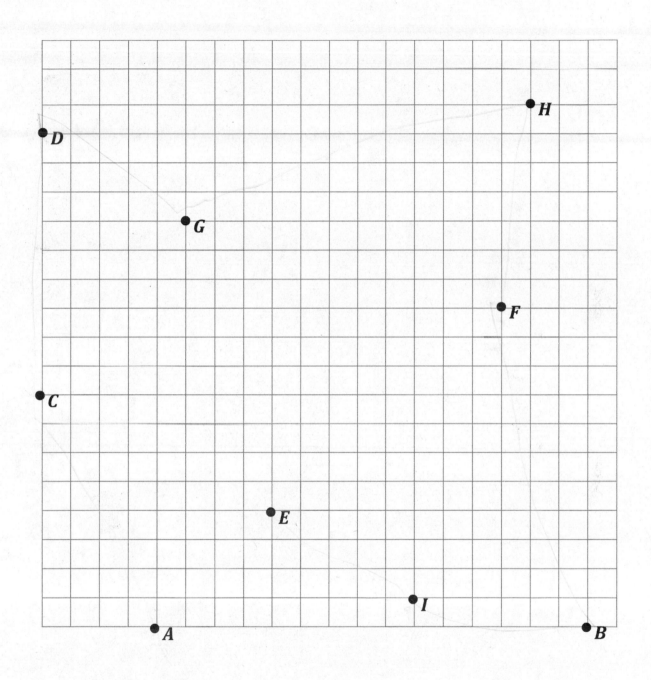

unlabeled coordinate plane

Lesson 3:   Name points using coordinate pairs, and use the coordinate pairs to
plot points.

Violet and Magnolia are shopping for boxes to organize the materials for their design company. Magnolia wants to get small boxes, which measure 16 in × 10 in × 7 in. Violet wants to get large boxes, which measure 32 in × 20 in × 14 in. How many small boxes will equal the volume of four large boxes?

_____

_____

_____

_____

Read   **Draw**   **Write**

Lesson 4:   Name points using coordinate pairs, and use the coordinate pairs to plot points.

27

© 2018 Great Minds®. eureka-math.org

Battleship Rules

Goal: To sink all of your opponent's ships by correctly guessing their coordinates.

Materials

- 1 grid sheet (per person/per game)
- Red crayon/marker for hits
- Black crayon/marker for misses
- Folder to place between players

Ships

- Each player must mark 5 ships on the grid.
  - Aircraft carrier—plot 5 points.
  - Battleship—plot 4 points.
  - Cruiser—plot 3 points.
  - Submarine—plot 3 points.
  - Patrol boat—plot 2 points.

Setup

- With your opponent, choose a unit length and fractional unit for the coordinate plane.
- Label the chosen units on both grid sheets.
- Secretly select locations for each of the 5 ships on your My Ships grid.
  - All ships must be placed horizontally or vertically on the coordinate plane.
  - Ships can touch each other, but they may not occupy the same coordinate.

Play

- Players take turns firing one shot to attack enemy ships.
- On your turn, call out the coordinates of your attacking shot. Record the coordinates of each attack shot.
- Your opponent checks his/her My Ships grid. If that coordinate is unoccupied, your opponent says, "Miss." If you named a coordinate occupied by a ship, your opponent says, "Hit."
- Mark each attempted shot on your Enemy Ships grid. Mark a black ✖ on the coordinate if your opponent says, "Miss." Mark a red ✓ on the coordinate if your opponent says, "Hit."
- On your opponent's turn, if he/she hits one of your ships, mark a red ✓ on that coordinate of your My Ships grid. When one of your ships has every coordinate marked with a ✓, say, "You've sunk my [name of ship]."

Victory

- The first player to sink all (or the most) opposing ships, wins.

Lesson 4:   Name points using coordinate pairs, and use the coordinate pairs to plot points.

© 2018 Great Minds®. eureka-math.org   29

My Ships

- Draw a red ✓ over any coordinate your opponent hits.
- Once all of the coordinates of any ship have been hit, say, "You've sunk my [name of ship]."

> Aircraft carrier—5 points
> Battleship—4 points
> Cruiser—3 points
> Submarine—3 points
> Patrol boat—2 points

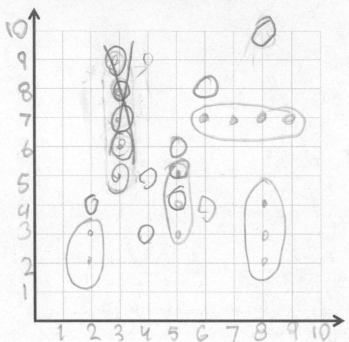

Enemy Ships

- Draw a black ✖ on the coordinate if your opponent says, "Miss."
- Draw a red ✓ on the coordinate if your opponent says, "Hit."
- Draw a circle around the coordinates of a sunken ship.

Attack Shots

- Record the coordinates of each shot below and whether it was a ✓ (hit) or an ✖ (miss).

(5 , 7) ✓   (6 , 4) ✓
(5 , 6) ✗   (8 , 4) ✗
(4 , 7) ✗   (8 , 2) ✗
(5 , 9) ✓   (5 , 2) ✗
(5 , 8) ✓   (7 , 5) ✓
(5 , 10) ✓   (7 , 4) ✓
(3 , 4) ✗   (7 , 3) ✓
(5 , 5) ✗   (7 , 2) ✓

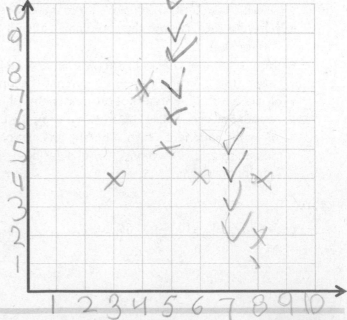

Name _____ Date _____

Fatima and Rihana are playing Battleship. They labeled their axes using just whole numbers.

a. Fatima's first guess is (2, 2). Rihana says, "Hit!" Give the coordinates of four points that Fatima might guess next.

b. Rihana says, "Hit!" for the points directly above and below (2, 2). What are the coordinates that Fatima guessed?

Lesson 4: Name points using coordinate pairs, and use the coordinate pairs to plot points.

31

A company has developed a new game. Cartons are needed to ship 40 games at a time. Each game is 2 inches high by 7 inches wide by 14 inches long.

How would you recommend packing the board games in the carton? What are the dimensions of a carton that could ship 40 board games with no extra room in the box?

_____

_____

_____

_____

Read   **Draw**   **Write**

Lesson 5:   Investigate patterns in vertical and horizontal lines, and interpret
             points on the plane as distances from the axes.

© 2018 Great Minds®. eureka-math.org

33

Name _____   Date _____

1. Use the coordinate plane to the right to answer the
    following questions.

    a. Use a straightedge to construct a line that goes
        through points A and B. Label the line e.

    b. Line e is parallel to the _____-axis and is
        perpendicular to the _____-axis.

    c. Plot two more points on line e. Name them
        C and D.

    d. Give the coordinates of each point below.

            A: _____ B: _____

            C: _____ D: _____

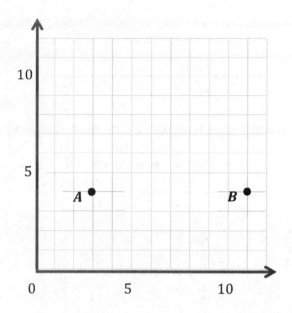

    e. What do all of the points of line e have in common?

    f. Give the coordinates of another point that would fall on line e with an x-coordinate greater than 15.

EUREKA
MATH

Lesson 5:   Investigate patterns in vertical and horizontal lines, and interpret
             points on the plane as distances from the axes.

© 2018 Great Minds®. eureka-math.org

35

2. Plot the following points on the coordinate plane to the right.

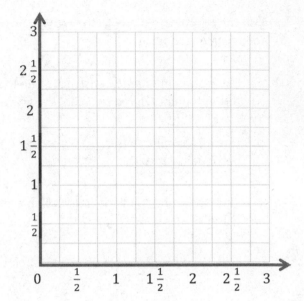

$P: (1\frac{1}{2}, \frac{1}{2})$ $Q: (1\frac{1}{2}, 2\frac{1}{2})$

$R: (1\frac{1}{2}, 1\frac{1}{4})$ $S: (1\frac{1}{2}, \frac{3}{4})$

a. Use a straightedge to draw a line to connect these points. Label the line h.

b. In line h, $x =$ _____ for all values of y.

c. Circle the correct word.

Line h is *parallel* *perpendicular* to the x-axis.

Line h is *parallel* *perpendicular* to the y-axis.

d. What pattern occurs in the coordinate pairs that let you know that line h is vertical?

3. For each pair of points below, think about the line that joins them. For which pairs is the line parallel to the x-axis? Circle your answer(s). Without plotting them, explain how you know.

a. (1.4, 2.2) and (4.1, 2.4)   b. (3, 9) and (8, 9)   c. $(1\frac{1}{4}, 2)$ and $(1\frac{1}{4}, 8)$

4. For each pair of points below, think about the line that joins them. For which pairs is the line parallel to the y-axis? Circle your answer(s). Then, give 2 other coordinate pairs that would also fall on this line.

a. (4, 12) and (6, 12)   b. $(\frac{3}{5}, 2\frac{3}{5})$ and $(\frac{1}{5}, 3\frac{1}{5})$ c. (0.8, 1.9) and (0.8, 2.3)

Lesson 5:   Investigate patterns in vertical and horizontal lines, and interpret points on the plane as distances from the axes.   EUREKA MATH

5. Write the coordinate pairs of 3 points that can be connected to construct a line that is $5\frac{1}{2}$ units to the right of and parallel to the y-axis.

    a. _____   b. _____   c. _____

6. Write the coordinate pairs of 3 points that lie on the x-axis.

    a. _____   b. _____   c. _____

7. Adam and Janice are playing Battleship. Presented in the table is a record of Adam's guesses so far.
    He has hit Janice's battleship using these coordinate pairs. What should he guess next? How do you know? Explain using words and pictures.

(3, 11)	hit
(2, 11)	miss
(3, 10)	hit
(4, 11)	miss
(3, 9)	miss

Lesson 5:   Investigate patterns in vertical and horizontal lines, and interpret
              points on the plane as distances from the axes.   37

© 2018 Great Minds®. eureka-math.org

Name _____   Date _____

1. Use a straightedge to construct a line that goes
    through points A and B. Label the line ℓ.

2. Which axis is parallel to line ℓ?

    Which axis is perpendicular to line ℓ?

3. Plot two more points on line ℓ. Name them C and D.

4. Give the coordinates of each point below.

    A: _____ B: _____

    C: _____ D: _____

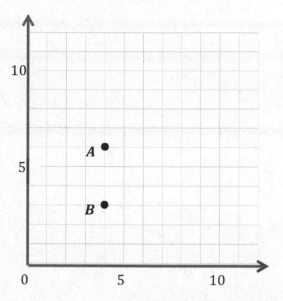

5. Give the coordinates of another point that falls on line ℓ with a y-coordinate greater than 20.

EUREKA
MATH

Lesson 5:   Investigate patterns in vertical and horizontal lines, and interpret
              points on the plane as distances from the axes.

39

© 2018 Great Minds®. eureka-math.org

Point	x	y	(x, y)
H			
I			
J			
K			
L			

a.

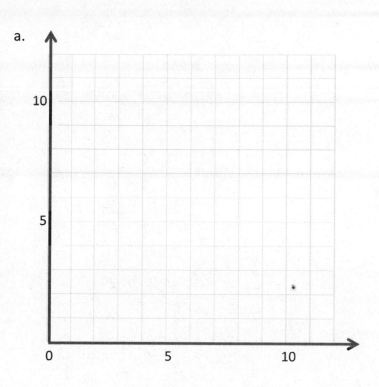

b.

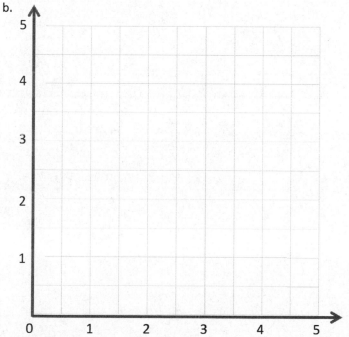

Point	x	y	(x, y)
D	$2\frac{1}{2}$	0	$(2\frac{1}{2}, 0)$
E	$2\frac{1}{2}$	2	$(2\frac{1}{2}, 2)$
F	$2\frac{1}{2}$	4	$(2\frac{1}{2}, 4)$

coordinate plane practice

Lesson 5:   Investigate patterns in vertical and horizontal lines, and interpret points on the plane as distances from the axes.

41

Adam built a toy box for his children's wooden blocks.

a. If the inside dimensions of the box are 18 inches by 12 inches by 6 inches, what is the maximum number of 2-inch wooden cubes that will fit in the toy box?

b. What if Adam had built the box 16 inches by 9 inches by 9 inches? What is the maximum number of 2-inch wooden cubes that would fit in this size box?

Read   **Draw**   **Write**

Lesson 6:   Investigate patterns in vertical and horizontal lines, and interpret points on the plane as distances from the axes.

© 2018 Great Minds®. eureka-math.org

Name _____   Date _____

1. Plot the following points, and label them on the coordinate plane.

    A: (0.3, 0.1)   B: (0.3, 0.7)

    C: (0.2, 0.9)   D: (0.4, 0.9)

    a. Use a straightedge to construct line segments
        $\overline{AB}$ and $\overline{CD}$.

    b. Line segment _____ is parallel to the
        x-axis and is perpendicular to the y-axis.

    c. Line segment _____ is parallel to the
        y-axis and is perpendicular to the x-axis.

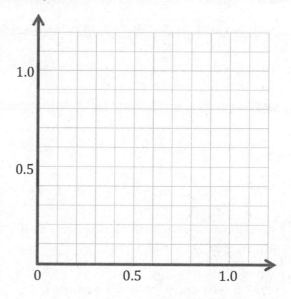

    d. Plot a point on line segment $\overline{AB}$ that is not at the endpoints, and name it U. Write the coordinates.
        U (_____, _____)

    e. Plot a point on line segment $\overline{CD}$, and name it V. Write the coordinates. V (_____, _____)

Lesson 6:   Investigate patterns in vertical and horizontal lines, and interpret
             points on the plane as distances from the axes.

© 2018 Great Minds®. eureka-math.org

45

2. Construct line f such that the y-coordinate of every point is $3\frac{1}{2}$, and construct line g such that the x-coordinate of every point is $4\frac{1}{2}$.

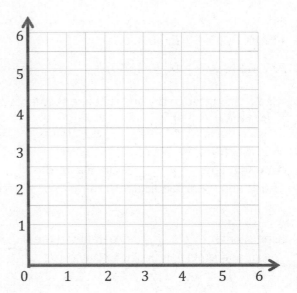

a. Line f is _____ units from the x-axis.

b. Give the coordinates of the point on line f that is $\frac{1}{2}$ unit from the y-axis. _____

c. With a blue pencil, shade the portion of the grid that is less than $3\frac{1}{2}$ units from the x-axis.

d. Line g is _____ units from the y-axis.

e. Give the coordinates of the point on line g that is 5 units from the x-axis. _____

f. With a red pencil, shade the portion of the grid that is more than $4\frac{1}{2}$ units from the y-axis.

46   Lesson 6:   Investigate patterns in vertical and horizontal lines, and interpret
                             points on the plane as distances from the axes.

                   © 2018 Great Minds®. eureka-math.org   EUREKA
                                                                                           MATH

3. Complete the following tasks on the plane below.

    a. Construct a line m that is perpendicular to the x-axis and 3.2 units from the y-axis.

    b. Construct a line a that is 0.8 unit from the x-axis.

    c. Construct a line t that is parallel to line m and is halfway between line m and the y-axis.

    d. Construct a line h that is perpendicular to line t and passes through the point (1.2, 2.4).

    e. Using a blue pencil, shade the region that contains points that are more than 1.6 units and less than 3.2 units from the y-axis.

    f. Using a red pencil, shade the region that contains points that are more than 0.8 unit and less than 2.4 units from the x-axis.

    g. Give the coordinates of a point that lies in the double-shaded region.

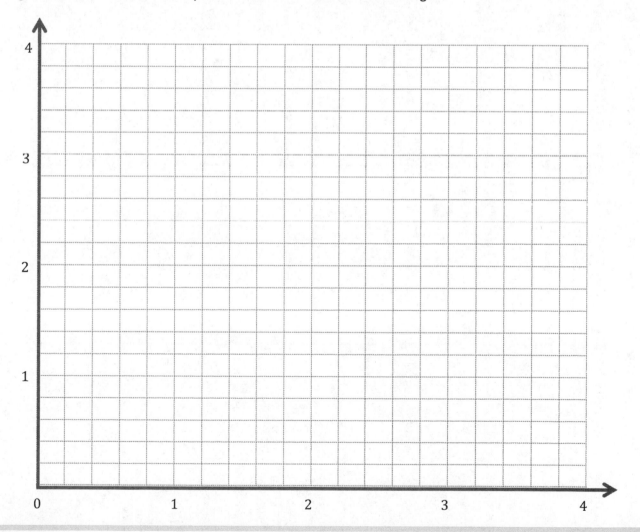

EUREKA
MATH

Lesson 6: Investigate patterns in vertical and horizontal lines, and interpret points on the plane as distances from the axes.

47

© 2018 Great Minds®. eureka-math.org

Name _____   Date _____

1. Plot the point H $(2\frac{1}{2}, 1\frac{1}{2})$.

2. Line ℓ passes through point H and is parallel to the y-axis. Construct line ℓ.

3. Construct line m such that the y-coordinate of every point is $\frac{3}{4}$.

4. Line m is _____ units from the x-axis.

5. Give the coordinates of the point on line m that is $\frac{1}{2}$ unit from the y-axis.

6. With a blue pencil, shade the portion of the plane that is less than $\frac{3}{4}$ unit from the x-axis.

7. With a red pencil, shade the portion of the plane that is less than $2\frac{1}{2}$ units from the y-axis.

8. Plot a point that lies in the double-shaded region. Give the coordinates of the point.

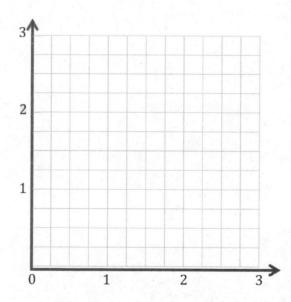

EUREKA
MATH®

Lesson 6:   Investigate patterns in vertical and horizontal lines, and interpret
             points on the plane as distances from the axes.

49

© 2018 Great Minds®. eureka-math.org

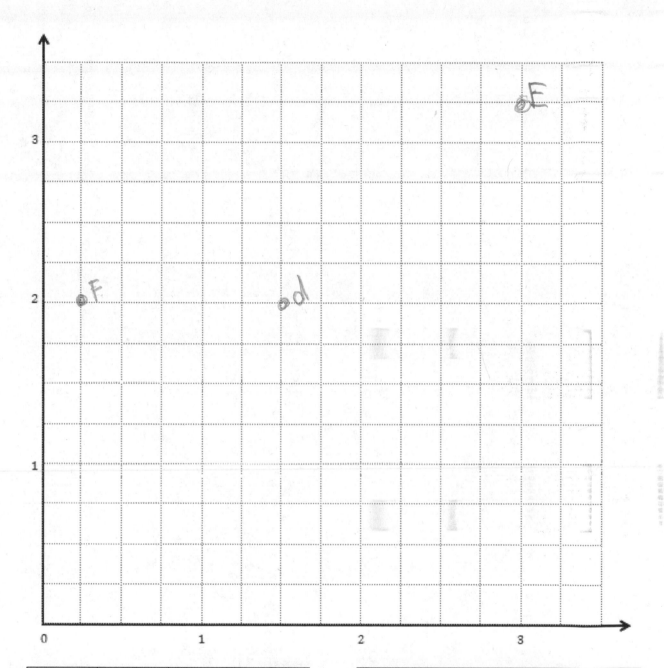

Point	x	y	(x, y)
A	2	$1\frac{1}{4}$	$(2, 1\frac{1}{4})$
B	$\frac{2}{4}$	$2\frac{3}{4}$	$(\frac{2}{4}, 2\frac{3}{4})$
C	3	$\frac{3}{4}$	$(3, \frac{3}{4})$

Point	x	y	(x, y)
D	$1\frac{2}{4}$	2	
E			
F			

coordinate plane

Lesson 6: Investigate patterns in vertical and horizontal lines, and interpret points on the plane as distances from the axes.

51

© 2018 Great Minds®. eureka-math.org

An orchard charges $0.85 to ship a quarter kilogram of grapefruit. Each grapefruit weighs approximately 165 grams. How much will it cost to ship 40 grapefruits?

_____

_____

_____

_____

Read   Draw   Write

Name _____ Date _____

1. Complete the chart. Then, plot the points on the coordinate plane below.

x	y	(x, y)
0	1	(0, 1)
2	3	
4	5	
6	7	

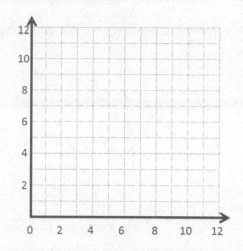

a. Use a straightedge to draw a line connecting these points.

b. Write a rule showing the relationship between the x- and y-coordinates of points on the line.

c. Name 2 other points that are on this line. _____   _____

2. Complete the chart. Then, plot the points on the coordinate plane below.

x	y	(x, y)
$\frac{1}{2}$	1	
1	2	
$1\frac{1}{2}$	3	
2	4	

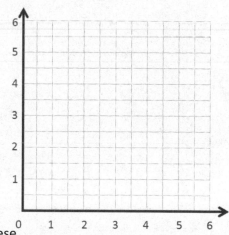

a. Use a straightedge to draw a line connecting these points.

b. Write a rule showing the relationship between the x- and y-coordinates.

c. Name 2 other points that are on this line. _____   _____

EUREKA
MATH

Lesson 7: Plot points, use them to draw lines in the plane, and describe patterns within the coordinate pairs.

55

© 2018 Great Minds®. eureka-math.org

3. Use the coordinate plane below to answer the following questions.

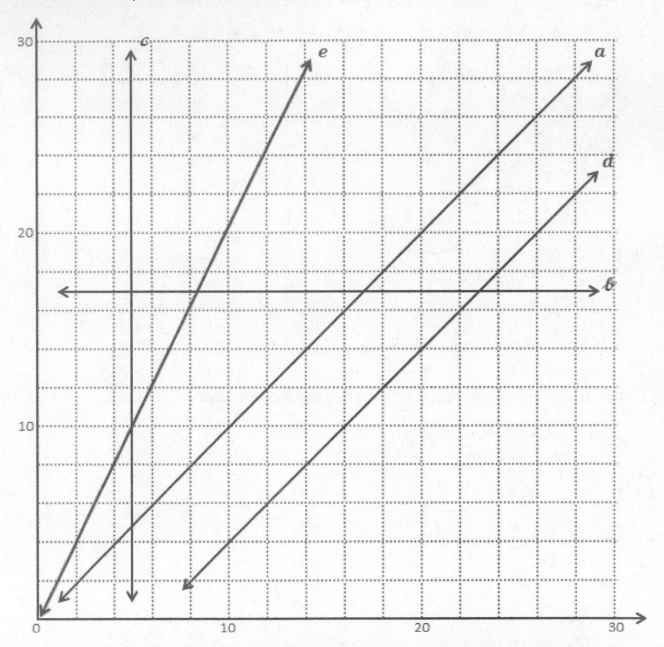

a. Give the coordinates for 3 points that are on line a. _____ _____ _____

b. Write a rule that describes the relationship between the x- and y-coordinates for the points on line a.

EUREKA
MATH

c. What do you notice about the *y*-coordinates of every point on line *b*?

d. Fill in the missing coordinates for points on line *d*.

(12, _____)   (6, _____)   (_____, 24)   (28, _____)   (_____, 28)

e. For any point on line *c*, the *x*-coordinate is _____.

f. Each of the points lies on at least 1 of the lines shown in the plane on the previous page. Identify a line that contains each of the following points.

i.   (7, 7) ___*a*___   ii.   (14, 8) _____   iii.   (5, 10) _____

iv.   (0, 17) _____   v.   (15.3, 9.3) _____   vi.   (20, 40) _____

EUREKA MATH   **Lesson 7:**   Plot points, use them to draw lines in the plane, and describe patterns   **57**
within the coordinate pairs.

© 2018 Great Minds®. eureka-math.org

Name _____   Date _____

Complete the chart. Then, plot the points on the coordinate plane.

x	y	(x, y)
0	4	
2	6	
3	7	
7	11	

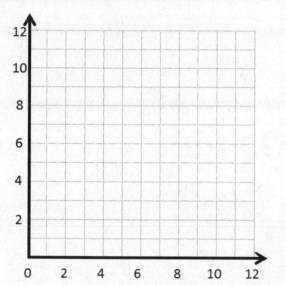

1. Use a straightedge to draw a line connecting these points.

2. Write a rule to show the relationship between the x- and y-coordinates for points on the line.

3. Name two other points that are also on this line.   _____   _____

EUREKA MATH

Lesson 7:   Plot points, use them to draw lines in the plane, and describe patterns within the coordinate pairs.

© 2018 Great Minds®. eureka-math.org

59

Name _____ Date _____

1.

a.

Point	x	y	(x, y)
A	0	0	(0, 0)
B	1	1	(1, 1)
C	2	2	(2, 2)
D	3	3	(3, 3)

b.

Point	x	y	(x, y)
G	0	3	(0, 3)
H	$\frac{1}{2}$	$3\frac{1}{2}$	$(\frac{1}{2}, 3\frac{1}{2})$
I	1	4	(1, 4)
J	$1\frac{1}{2}$	$4\frac{1}{2}$	$(1\frac{1}{2}, 4\frac{1}{2})$

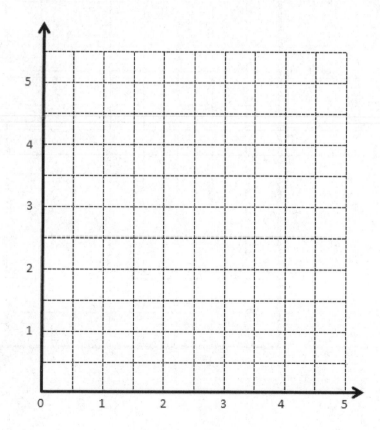

coordinate plane

EUREKA MATH

Lesson 7: Plot points, use them to draw lines in the plane, and describe patterns within the coordinate pairs.

61

© 2018 Great Minds®. eureka-math.org

2.

a

Point	(x, y)
L	(0, 3)
M	(2, 3)
N	(4, 3)

b.

Point	(x, y)
O	(0, 0)
P	(1, 2)
Q	(2, 4)

c.

Point	(x, y)
R	$(1, \frac{1}{2})$
S	$(2, 1\frac{1}{2})$
T	$(3, 2\frac{1}{2})$

d.

Point	(x, y)
U	(1, 3)
V	(2, 6)
W	(3, 9)

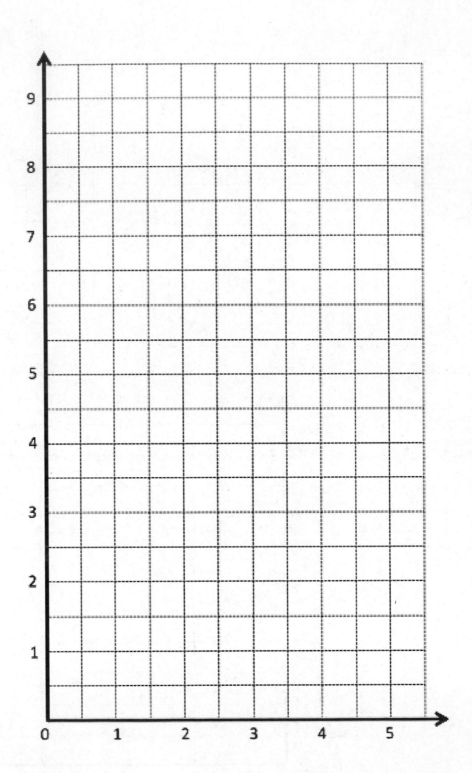

coordinate plane

Lesson 7:   Plot points, use them to draw lines in the plane, and describe patterns
within the coordinate pairs.

EUREKA
MATH

The coordinate pairs listed locate points on two different lines. Write a rule that describes the relationship between the x- and y-coordinates for each line.

Line ℓ: $(3\frac{1}{2}, 7)$, $(1\frac{2}{3}, 3\frac{1}{3})$, $(5, 10)$

Line m: $(\frac{6}{3}, 1)$, $(3\frac{1}{2}, 1\frac{3}{4})$, $(13, 6\frac{1}{2})$

_____

_____

_____

_____

Read   **Draw**   **Write**

Name _____   Date _____

1. Create a table of 3 values for x and y such that each y-coordinate is 3 more than the corresponding x-coordinate.

x	y	(x, y)

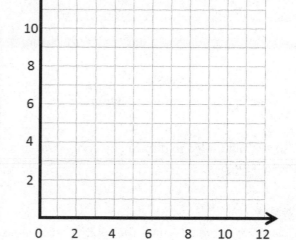

a. Plot each point on the coordinate plane.

b. Use a straightedge to draw a line connecting these points.

c. Give the coordinates of 2 other points that fall on this line with x-coordinates greater than 12.
   (_____, _____) and (_____, _____)

2. Create a table of 3 values for x and y such that each y-coordinate is 3 times as much as its corresponding x-coordinate.

x	y	(x, y)

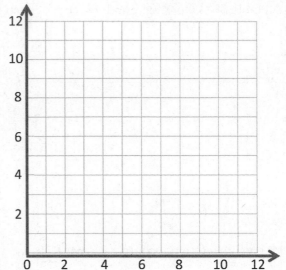

a. Plot each point on the coordinate plane.

b. Use a straightedge to draw a line connecting these points.

c. Give the coordinates of 2 other points that fall on this line with y-coordinates greater than 25.
(_____, _____) and (_____, _____)

66   Lesson 8:   Generate a number pattern from a given rule, and plot the points.

EUREKA MATH

3. Create a table of 5 values for x and y such that each y-coordinate is 1 more than 3 times as much as its corresponding x value.

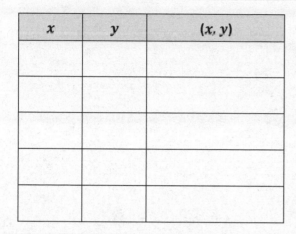

x	y	(x, y)

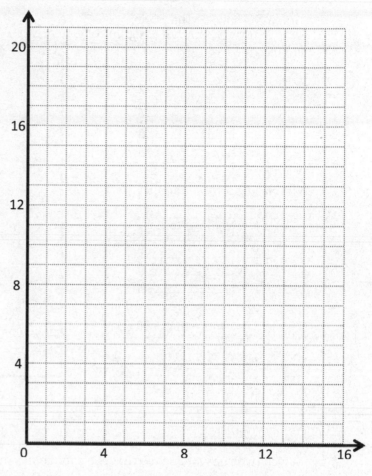

a. Plot each point on the coordinate plane.

b. Use a straightedge to draw a line connecting these points.

c. Give the coordinates of 2 other points that would fall on this line whose x-coordinates are greater than 12.

    (_____, _____) and (_____, _____)

4. Use the coordinate plane below to complete the following tasks.

   a. Graph the lines on the plane.

      line ℓ: x is equal to y

	x	y	(x, y)
A			
B			
C			

      line m: y is 1 more than x

	x	y	(x, y)
G			
H			
I			

      line n: y is 1 more than twice x

	x	y	(x, y)
S			
T			
U			

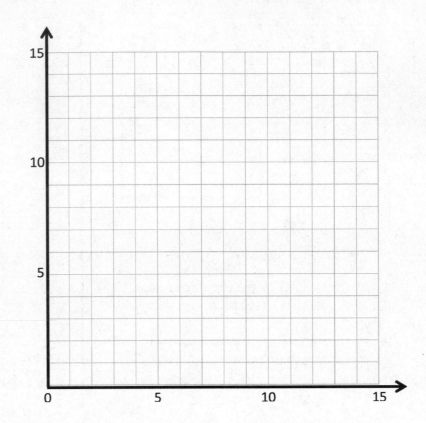

   b. Which two lines intersect? Give the coordinates of their intersection.

   c. Which two lines are parallel?

   d. Give the rule for another line that would be parallel to the lines you listed in Problem 4(c).

Lesson 8:   Generate a number pattern from a given rule, and plot the points.

EUREKA
MATH

Name _____   Date _____

Complete this table with values for y such that each y-coordinate is 5 more than 2 times as much as its corresponding x-coordinate.

x	y	(x, y)
0		
2		
3.5		

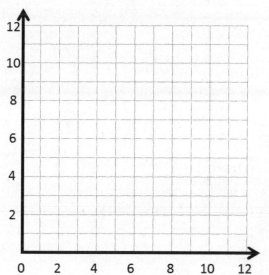

a. Plot each point on the coordinate plane.

b. Use a straightedge to draw a line connecting these points.

c. Name 2 other points that fall on this line with y-coordinates greater than 25.

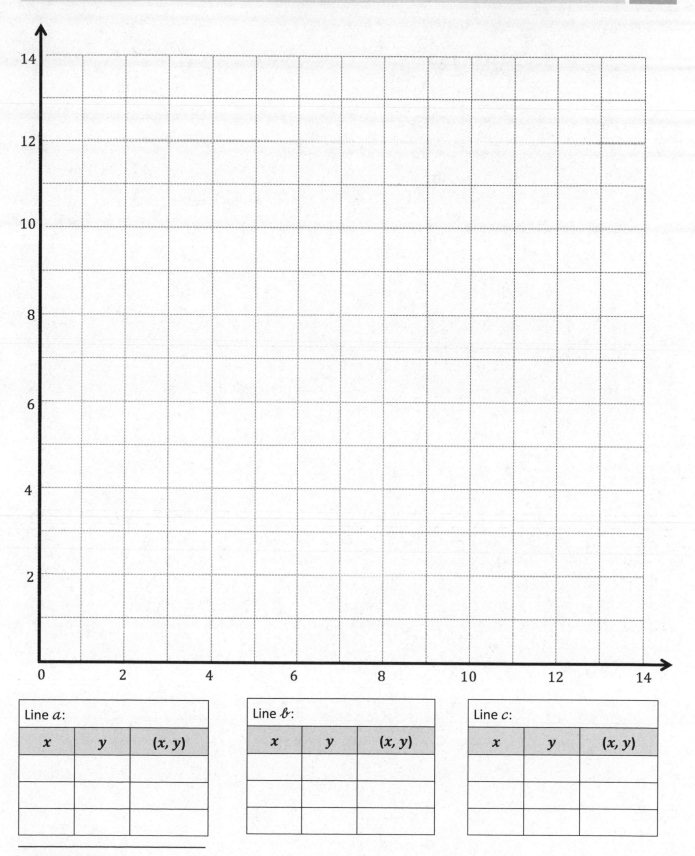

Line *a*:		
x	*y*	*(x, y)*

Line *b*:		
x	*y*	*(x, y)*

Line *c*:		
x	*y*	*(x, y)*

coordinate plane

© 2018 Great Minds®. eureka-math.org

Maggie spent $46.20 to buy pencil sharpeners for her gift shop. If each pencil sharpener costs 60 cents, how many pencil sharpeners did she buy? Solve by using the standard algorithm.

Read   **Draw**   **Write**

Lesson 9:   Generate two number patterns from given rules, plot the points, and analyze the patterns.

© 2018 Great Minds®. eureka-math.org

73

Name _____   Date _____

1. Complete the table for the given rules.

    Line a

    Rule: y is 1 more than x

x	y	(x, y)
1		
5		
9		
13		

    Line b

    Rule: y is 4 more than x

x	y	(x, y)
0		
5		
8		
11		

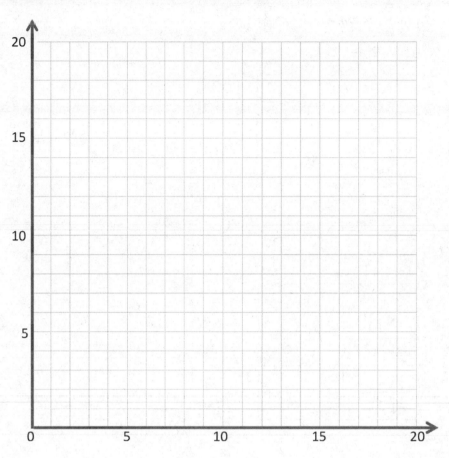

a. Construct each line on the coordinate plane above.

b. Compare and contrast these lines.

c. Based on the patterns you see, predict what line c, whose rule is *y is 7 more than x*, would look like. Draw your prediction on the plane above.

2. Complete the table for the given rules.

Line *e*

Rule: y is twice as much as x

x	y	(x, y)
0		
2		
5		
9		

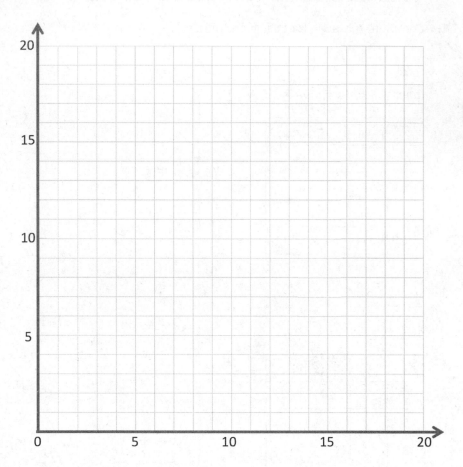

Line *f*

Rule: y is half as much as x

x	y	(x, y)
0		
6		
10		
20		

a. Construct each line on the coordinate plane above.

b. Compare and contrast these lines.

c. Based on the patterns you see, predict what line *g*, whose rule is *y is 4 times as much as x*, would look like. Draw your prediction in the plane above.

Lesson 9:   Generate two number patterns from given rules, plot the points, and
                              analyze the patterns.

EUREKA
MATH®

Name _____   Date _____

Complete the table for the given rules. Then, construct lines ℓ and m on the coordinate plane.

Line ℓ

Rule: y is 5 more than x

x	y	(x, y)
0		
1		
2		
4		

Line m

Rule: y is 5 times as much as x

x	y	(x, y)
0		
1		
2		
4		

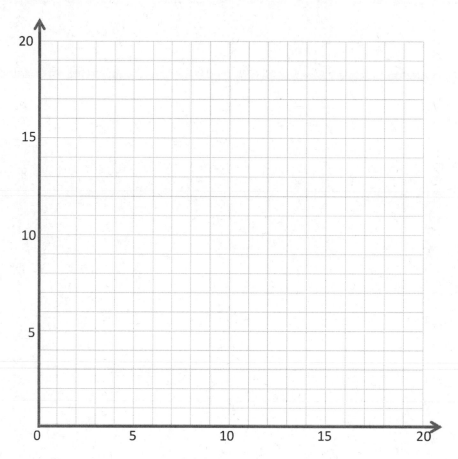

EUREKA
MATH

Lesson 9:   Generate two number patterns from given rules, plot the points, and
            analyze the patterns.

77

© 2018 Great Minds®. eureka-math.org

Line ℓ

Rule: *y is 2 more than x*

x	y	(x, y)
1		
5		
10		
15		

Line m

Rule: *y is 5 more than x*

x	y	(x, y)
0		
5		
10		
15		

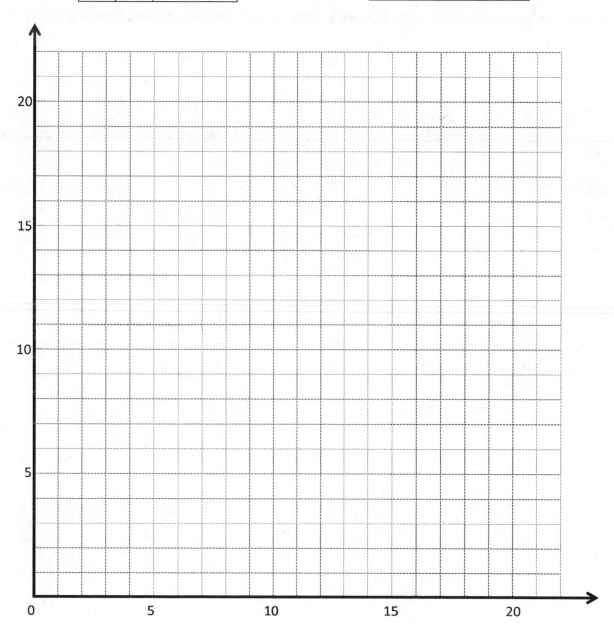

coordinate plane

Lesson 9:   Generate two number patterns from given rules, plot the points, and
             analyze the patterns.

79

© 2018 Great Minds®. eureka-math.org

Line *p*

Rule: y is x times 2

x	*y*	*(x, y)*

Line *q*

Rule: y is x times 3

x	*y*	*(x, y)*

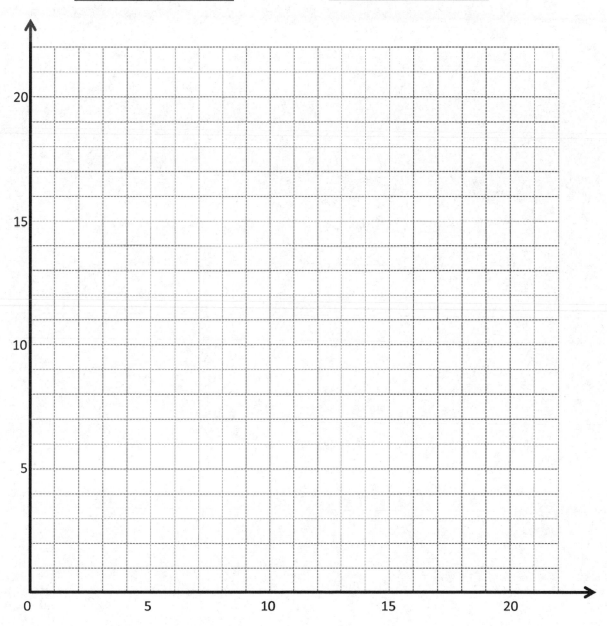

coordinate plane

Lesson 9:   Generate two number patterns from given rules, plot the points, and analyze the patterns.

81

A 12-man relay team runs a 45 km race. Each member of the team runs an equal distance. How many kilometers does each team member run? One lap around the track is 0.75 km. How many laps does each team member run during the race?

_____

_____

_____

_____

Read   **Draw**   **Write**

Lesson 10:   Compare the lines and patterns generated by addition rules and
multiplicative rules.

© 2018 Great Minds®. eureka-math.org

83

Name _____   Date _____

1. Use the coordinate plane below to complete the following tasks.

    a. Line p represents the rule *x and y are equal*.

    b. Construct a line, d, that is parallel to line p and contains point D.

    c. Name 3 coordinate pairs on line d.

    d. Identify a rule to describe line d.

    e. Construct a line, e, that is parallel to line p and contains point E.

    f. Name 3 points on line e.

    g. Identify a rule to describe line e.

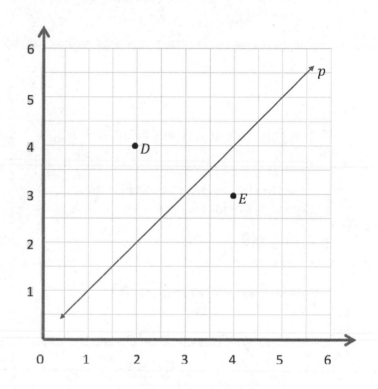

    h. Compare and contrast lines d and e in terms of their relationship to line p.

2. Write a rule for a fourth line that would be parallel to those above and would contain the point $(3\frac{1}{2}, 6)$. Explain how you know.

EUREKA MATH®

Lesson 10:   Compare the lines and patterns generated by addition rules and multiplicative rules.

© 2018 Great Minds®. eureka-math.org

85

3. Use the coordinate plane below to complete the following tasks.

    a. Line p represents the rule *x and y are equal*.

    b. Construct a line, v, that contains the origin and point V.

    c. Name 3 points on line v.

    d. Identify a rule to describe line v.

    e. Construct a line, w, that contains the origin and point W.

    f. Name 3 points on line w.

    g. Identify a rule to describe line w.

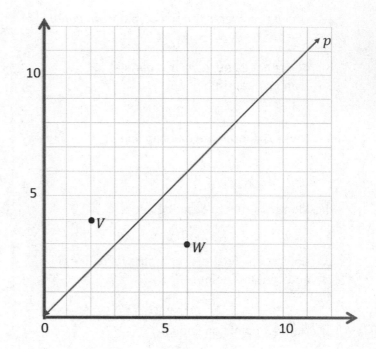

    h. Compare and contrast lines v and w in terms of their relationship to line p.

    i. What patterns do you see in lines that are generated by multiplication rules?

4. Circle the rules that generate lines that are parallel to each other.

    add 5 to x   *multiply x by $\frac{2}{3}$*   *x plus $\frac{1}{2}$*   *x times $1\frac{1}{2}$*

EUREKA
MATH®

Name _____   Date _____

Use the coordinate plane below to complete the following tasks.

a. Line *p* represents the rule *x and y are equal*.

b. Construct a line, *a*, that is parallel to line *p* and contains point *A*.

c. Name 3 points on line *a*.

d. Identify a rule to describe line *a*.

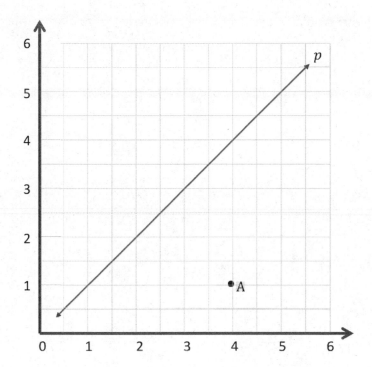

Lesson 10:   Compare the lines and patterns generated by addition rules and multiplicative rules.

© 2018 Great Minds®. eureka-math.org

87

Line p	Line b	Line c	Line d

Rule: *y is 0 more than x*   Rule: _____   Rule: _____   Rule: _____

x	y	(x, y)
0		
5		
10		
15		

x	y	(x, y)
7		
10		
13		
18		

x	y	(x, y)
2		
4		
8		
11		

x	y	(x, y)
5		
7		
12		
15		

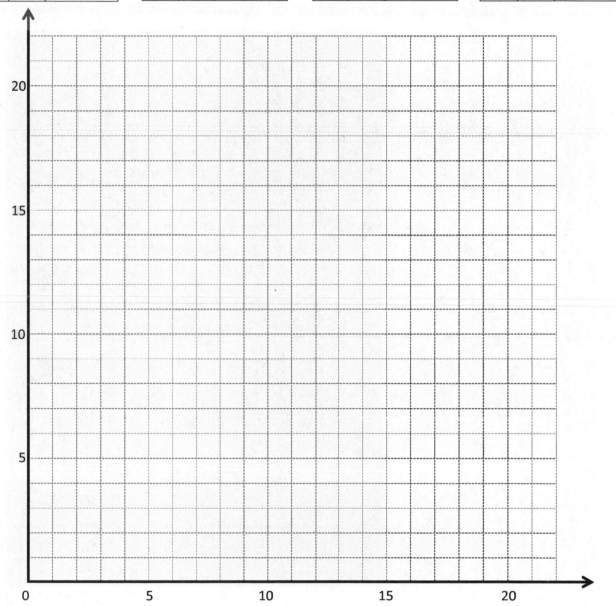

coordinate plane

Lesson 10:   Compare the lines and patterns generated by addition rules and
multiplicative rules.

© 2018 Great Minds®. eureka-math.org

89

Line g Rule: _____   *Line h Rule:* _____

x	y	(x, y)
1		
2		
5		
7		

x	y	(x, y)
3		
6		
12		
15		

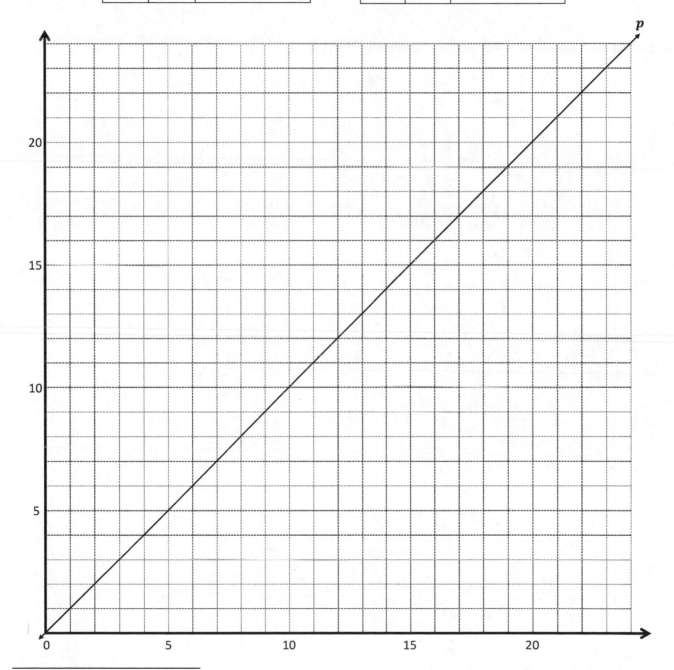

coordinate plane

Lesson 10:   Compare the lines and patterns generated by addition rules and
multiplicative rules.

91

© 2018 Great Minds®. eureka-math.org

Michelle has 3 kg of strawberries that she divided equally into small bags with 15 kg in each bag.

   a. How many bags of strawberries did she make?

   b. She gave a bag to her friend, Sarah. Sarah ate half of her strawberries. How many grams of strawbernes does Sarah have left?

Read   **Draw**   **Write**

Name _____   Date _____

1. Complete the tables for the given rules.

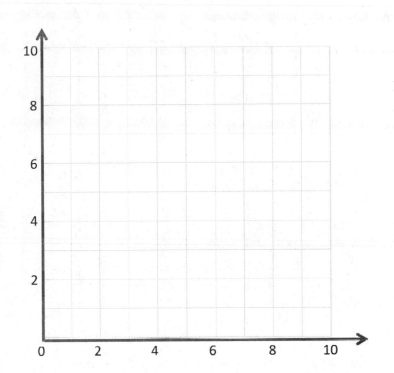

Line ℓ

Rule: *Double x*

x	y	(x, y)
0		
1		
2		
3		

Line *m*

Rule: *Double x, and then add 1*

x	y	(x, y)
0		
1		
2		
3		

a. Draw each line on the coordinate plane above.

b. Compare and contrast these lines.

c. Based on the patterns you see, predict what the line for the rule *double x, and then subtract 1* would look like. Draw the line on the plane above.

2. Circle the point(s) that the line for the rule *multiply x by $\frac{1}{3}$, and then add 1* would contain.

    $(0, \frac{1}{3})$ $(2, 1\frac{2}{3})$ $(1\frac{1}{2}, 1\frac{1}{2})$ $(2\frac{1}{4}, 2\frac{1}{4})$

a. Explain how you know.

b. Give two other points that fall on this line.

EUREKA
MATH®

© 2018 Great Minds®. eureka-math.org

3. Complete the tables for the given rules.

Line ℓ
Rule: *Halve x*

x	y	(x, y)
0		
1		
2		
3		

Line m
Rule: *Halve x, and then add 1 $\frac{1}{2}$*

x	y	(x, y)
0		
1		
2		
3		

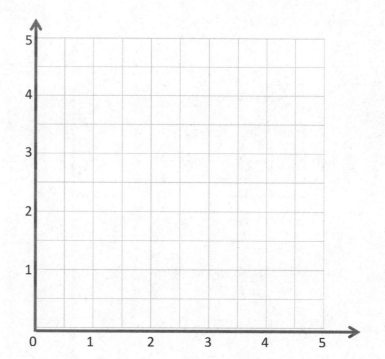

a. Draw each line on the coordinate plane above.

b. Compare and contrast these lines.

c. Based on the patterns you see, predict what the line for the rule *halve x, and then subtract 1* would look like. Draw the line on the plane above.

4. Circle the point(s) that the line for the rule *multiply x by $\frac{2}{3}$, and then subtract 1* would contain.

$(1\frac{1}{3}, \frac{1}{9})$ $(2, \frac{1}{3})$ $(1\frac{3}{2}, 1\frac{1}{2})$ $(3, 1)$

a. Explain how you know.

b. Give two other points that fall on this line.

EUREKA MATH

Name _____   Date _____

1. Complete the tables for the given rules.

Line ℓ

Rule: *Triple x*

x	y	(x, y)
0		
1		
2		
3		

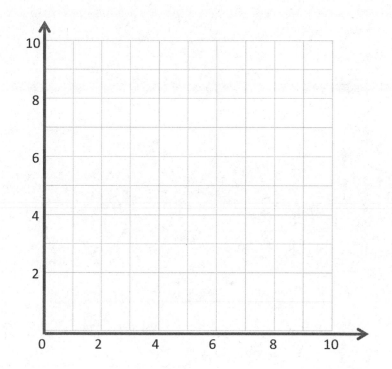

Line *m*

Rule: *Triple x, and then*

add 1

x	y	(x, y)
0		
1		
2		
3		

a. Draw each line on the coordinate plane above.

b. Compare and contrast these lines.

3. Circle the point(s) that the line for the rule *multiply x by $\frac{1}{3}$, and then add 1* would contain.

$(0, \frac{1}{2})$ $(1, 1\frac{1}{3})$ $(2, 1\frac{2}{3})$ $(3, 2\frac{1}{2})$

Line ℓ

Rule: *Triple x*

x	y	(x, y)
0		
1		
2		
4		

Line m

Rule: *Triple x, and then add 3*

x	y	(x, y)
0		
1		
2		
3		

Line n

Rule: *Triple x, and then subtract 2*

x	y	(x, y)
1		
2		
3		
4		

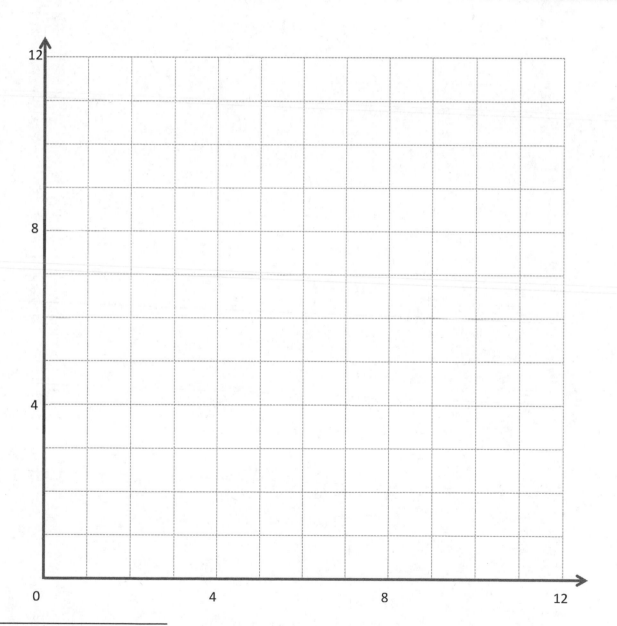

coordinate plane

EUREKA
MATH

Lesson 11:   Analyze number patterns created from mixed operations.

99

© 2018 Great Minds®. eureka-math.org

Mr. Jones had 640 books. He sold $\frac{1}{4}$ of them for \$2.00 each in the month of September. He sold half of the remaining books in October. Each book he sold in October earned $\frac{3}{4}$ of what each book sold for in September. How much money did Mr. Jones earn selling books? Show your thinking with a tape diagram.

Read   **Draw**   **Write**

EUREKA
MATH

Name _____   Date _____

1. Write a rule for the line that contains the points $(0, \frac{3}{4})$ and $(2\frac{1}{2}, 3\frac{1}{4})$.

    a. Identify 2 more points on this line. Draw the line on the grid below.

Point	x	y	(x, y)
B			
C			

    b. Write a rule for a line that is parallel to $\overleftrightarrow{BC}$ and goes through point $(1, \frac{1}{4})$.

2. Create a rule for the line that contains the points $(1, \frac{1}{4})$ and $(3, \frac{3}{4})$.

    a. Identify 2 more points on this line. Draw the line on the grid on the right.

Point	x	y	(x, y)
G			
H			

    b. Write a rule for a line that passes through the origin and lies between $\overleftrightarrow{BC}$ and $\overleftrightarrow{GH}$.

EUREKA
MATH

Lesson 12:   Create a rule to generate a number pattern, and plot the points.

103

3. Create a rule for a line that contains the point $(\frac{1}{4}, 1\frac{1}{4})$ using the operation or description below. Then, name 2 other points that would fall on each line.

    a. Addition: _____

Point	x	y	(x, y)
T			
U			

    b. A line parallel to the x-axis: _____

Point	x	y	(x, y)
G			
H			

    c. Multiplication: _____

Point	x	y	(x, y)
A			
B			

    d. A line parallel to the y-axis: _____

Point	x	y	(x, y)
V			
W			

    e. Multiplication with addition: _____

Point	x	y	(x, y)
R			
S			

4. Mrs. Boyd asked her students to give a rule that could describe a line that contains the point (0.6, 1.8). Avi said the rule could be *multiply x by 3*. Ezra claims this could be a vertical line, and the rule could be *x is always 0.6*. Erik thinks the rule could be *add 1.2 to x*. Mrs. Boyd says that all the lines they are describing could describe a line that contains the point she gave. Explain how that is possible, and draw the lines on the coordinate plane to support your response.

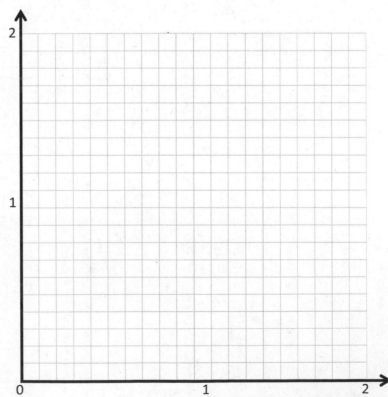

Lesson 12:   Create a rule to generate a number pattern, and plot the points.

Extension:

5. Create a mixed operation rule for the line that contains the points (0, 1) and (1, 3).

a. Identify 2 more points, O and P, on this line. Draw the line on the grid.

Point	x	y	(x, y)
O			
P			

b. Write a rule for a line that is parallel to $\overleftrightarrow{OP}$ and goes through point $(1, \, 2\frac{1}{2})$.

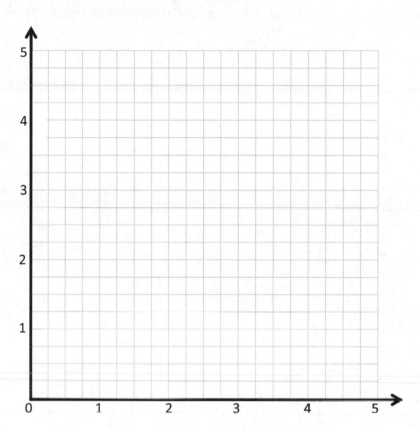

EUREKA
MATH®

Lesson 12:   Create a rule to generate a number pattern, and plot the points.

105

© 2018 Great Minds®. eureka-math.org

Name _____   Date _____

Write the rule for the line that contains the points $(0, 1\frac{1}{2})$ and $(1\frac{1}{2}, 3)$.

a. Identify 2 more points on this line.
    Draw the line on the grid.

Point	x	y	(x, y)
B			
C			

b. Write a rule for a line that is
    parallel to $\overleftrightarrow{BC}$ and goes through
    $(1, \frac{1}{2})$.

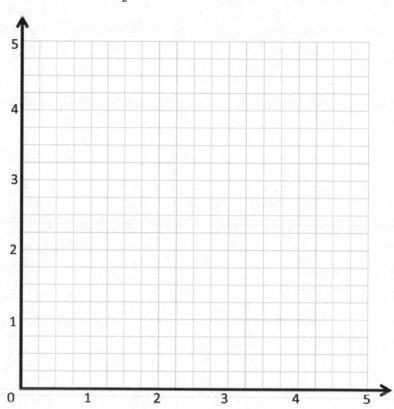

EUREKA
MATH

Lesson 12:   Create a rule to generate a number pattern, and plot the points.

107

© 2018 Great Minds®. eureka-math.org

Line l Line m

Rule: _____   Rule: _____

Point	x	y	(x, y)
A	$1\frac{1}{2}$	3	$(1\frac{1}{2}, 3)$
B			
C			
D			

Point	x	y	(x, y)
A			
E			
F			
G			

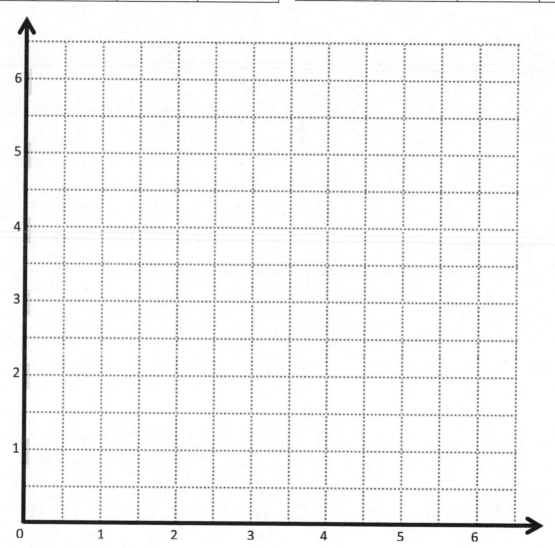

coordinate plane

Lesson 12:   Create a rule to generate a number pattern, and plot the points.   **109**

EUREKA MATH

© 2018 Great Minds®. eureka-math.org

Name _____ Date _____

1. Use a right angle template and straightedge to draw at least four sets of parallel lines in the space below.

2. Circle the segments that are parallel.

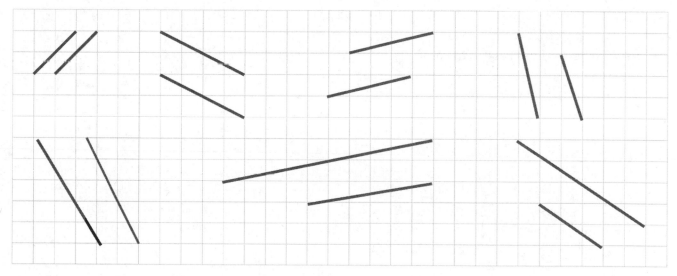

Lesson 13: Construct parallel line segments on a rectangular grid.

111

3. Use your straightedge to draw a segment parallel to each segment through the given point.

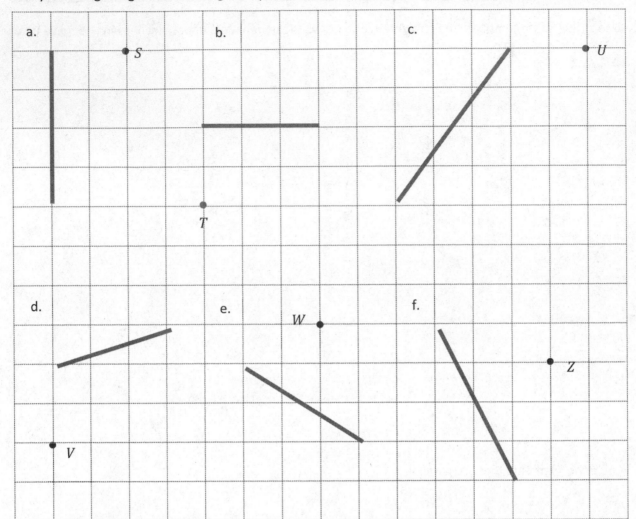

4. Draw 2 different lines parallel to line 𝒷.

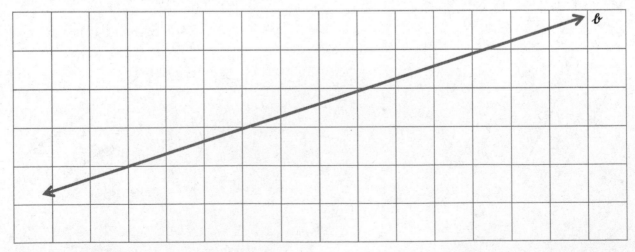

Lesson 13:   Construct parallel line segments on a rectangular grid.

EUREKA MATH

Name _____   Date _____

Use your straightedge to draw a segment parallel to each segment through the given point.

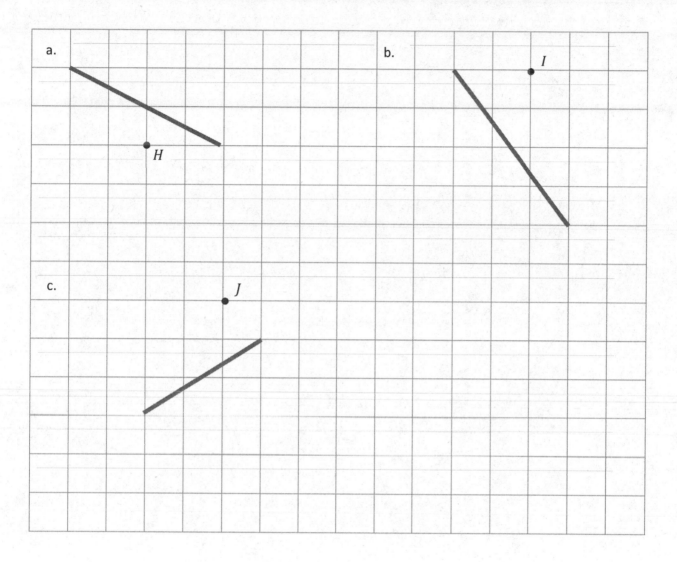

EUREKA
MATH®

a.↓   b.↓   c.↓   d.↓

e.→   f.↓   g.→   h.→

rectangles

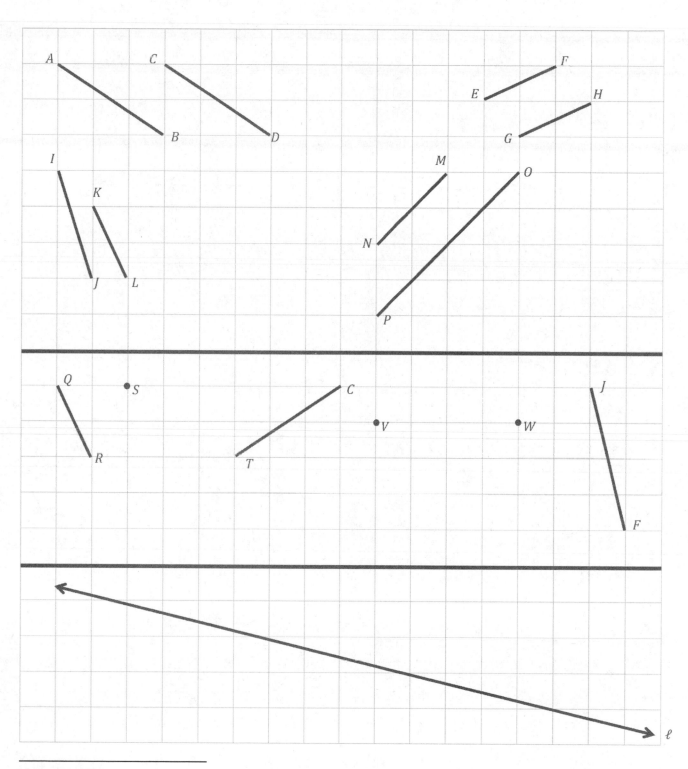

recording sheet

Drew's fish tank measures 32 cm by 22 cm by 26 cm. He pours 20 liters of water into it, and some water overflows the tank. Find the volume of water, in milliliters, that overflows.

Read   Draw   Write

Lesson 14:   Construct parallel line segments, and analyze relationships of the coordinate pairs.

119

© 2018 Great Minds®. eureka-math.org

Name _____ Date _____

1. Use the coordinate plane below to complete the following tasks.

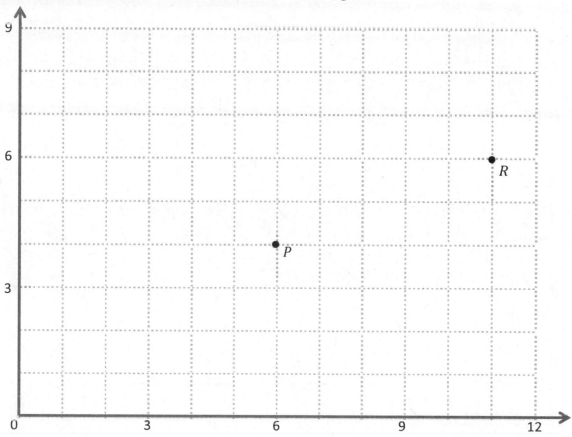

a. Identify the locations of P and R. P: (_____ , _____) R: (_____ , _____)

b. Draw $\overrightarrow{PR}$.

c. Plot the following coordinate pairs on the plane.

S: (6, 7) T: (11, 9)

d. Draw $\overrightarrow{ST}$.

e. Circle the relationship between $\overrightarrow{PR}$ and $\overrightarrow{ST}$. $\overrightarrow{PR} \perp \overrightarrow{ST}$ $\overrightarrow{PR} \| \overrightarrow{ST}$

f. Give the coordinates of a pair of points, U and V, such that $\overrightarrow{UV} \| \overrightarrow{PR}$.

U: (_____ , _____) V: (_____ , _____)

g. Draw $\overrightarrow{UV}$.

EUREKA MATH

Lesson 14:   Construct parallel line segments, and analyze relationships of the coordinate pairs.

121

© 2018 Great Minds®. eureka-math.org

2. Use the coordinate plane below to complete the following tasks.

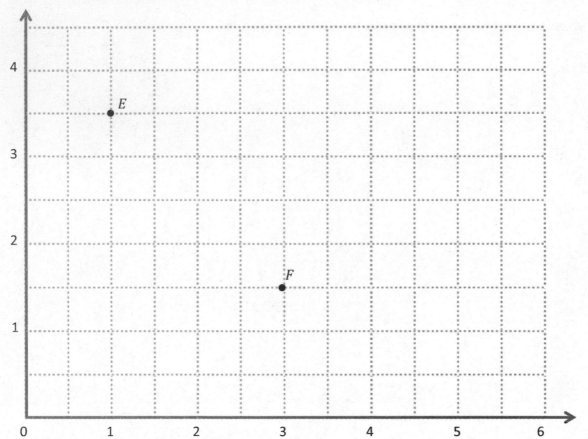

a. Identify the locations of E and F. $E:$ (_____, _____) $F:$ (_____, _____)

b. Draw $\overrightarrow{EF}$.

c. Generate coordinate pairs for L and M, such that $\overrightarrow{EF} \| \overrightarrow{LM}$.

$L:$ (____, ____) $M:$ (____, ____)

d. Draw $\overleftrightarrow{LM}$.

e. Explain the pattern you made use of when generating coordinate pairs for L and M.

f. Give the coordinates of a point, H, such that $\overrightarrow{EF} \| \overrightarrow{GH}$.

$G: \left(1\frac{1}{2}, 4\right)$ $H:$ (____, ____)

g. Explain how you chose the coordinates for H.

Lesson 14:   Construct parallel line segments, and analyze relationships of the
coordinate pairs.

EUREKA
MATH®

Name _____   Date _____

Use the coordinate plane below to complete the following tasks.

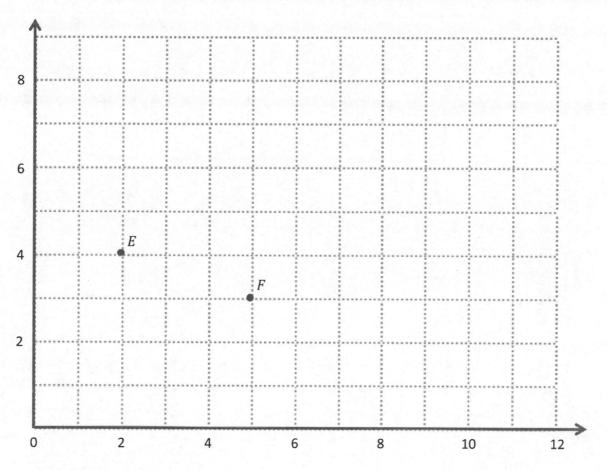

a. Identify the locations of *E* and *F*.   *E*: (_____, _____)   *F*: (_____, _____)

b. Draw $\overleftrightarrow{EF}$.

c. Generate coordinate pairs for *L* and *M*, such that $\overleftrightarrow{EF} \parallel \overleftrightarrow{LM}$.

   L: (____, ____)   *M*: (____, ____)

d. Draw $\overleftrightarrow{LM}$.

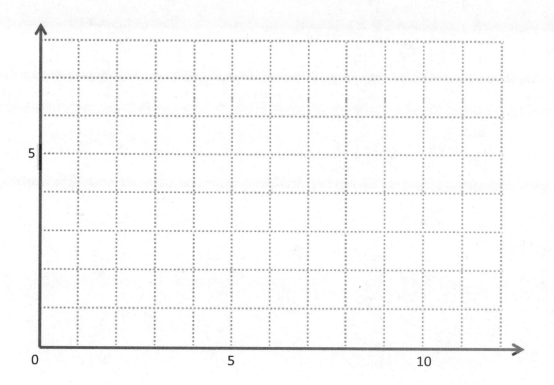

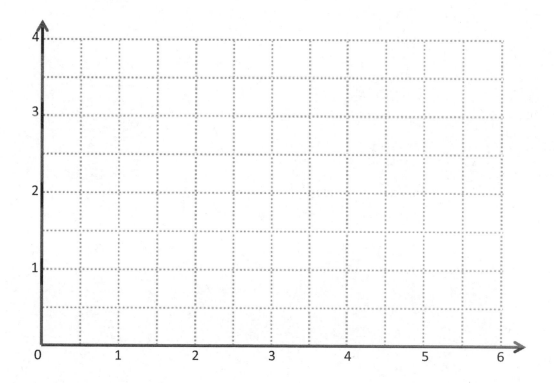

coordinate plane

EUREKA
MATH

Lesson 14:   Construct parallel line segments, and analyze relationships of the
coordinate pairs.

125

© 2018 Great Minds®. eureka-math.org

Name _____ Date _____

1. Circle the pairs of segments that are perpendicular.

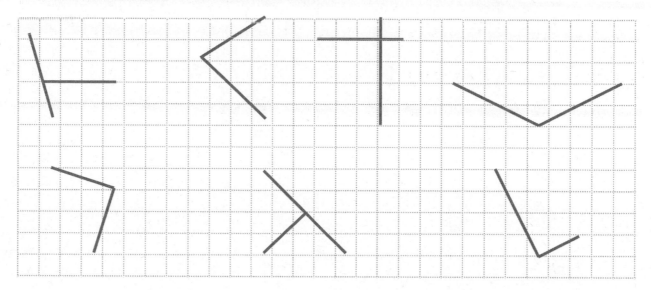

2. In the space below, use your right triangle templates to draw at least 3 different sets of perpendicular lines.

EUREKA MATH

Lesson 15: Construct perpendicular line segments on a rectangular grid.

127

© 2018 Great Minds®. eureka-math.org

3. Draw a segment perpendicular to each given segment. Show your thinking by sketching triangles as needed.

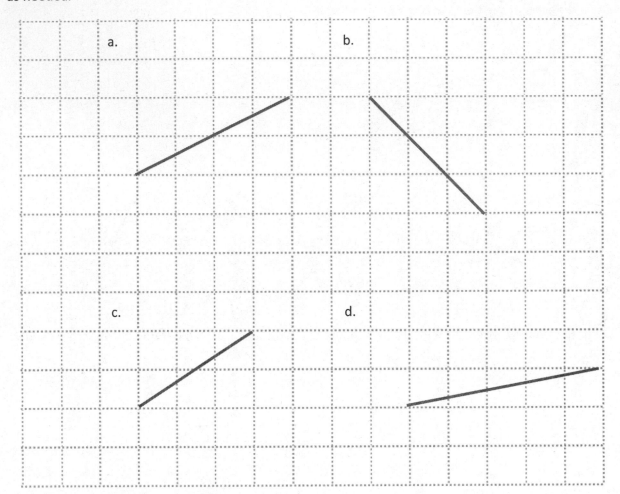

a.

b.

c.

d.

4. Draw 2 different lines perpendicular to line *e*.

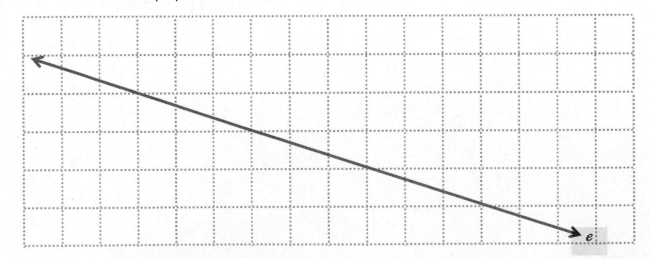

e

EUREKA
MATH

Name _____   Date _____

Draw a segment perpendicular to each given segment. Show your thinking by sketching triangles as needed.

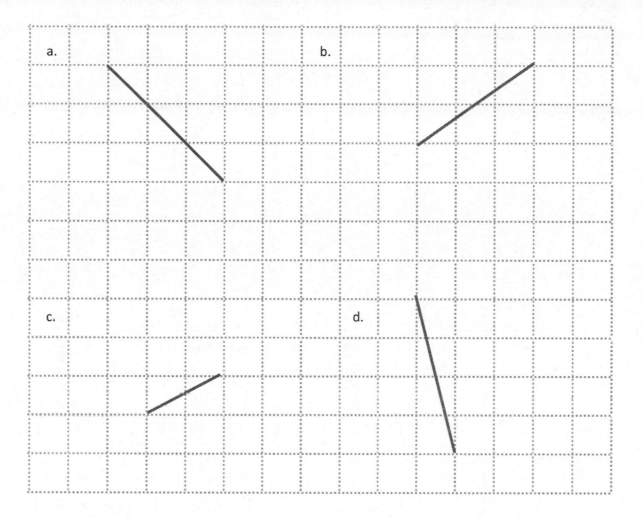

a.

b.

c.

d.

EUREKA MATH

Lesson 15:   Construct perpendicular line segments on a rectangular grid.

129

© 2018 Great Minds®. eureka-math.org

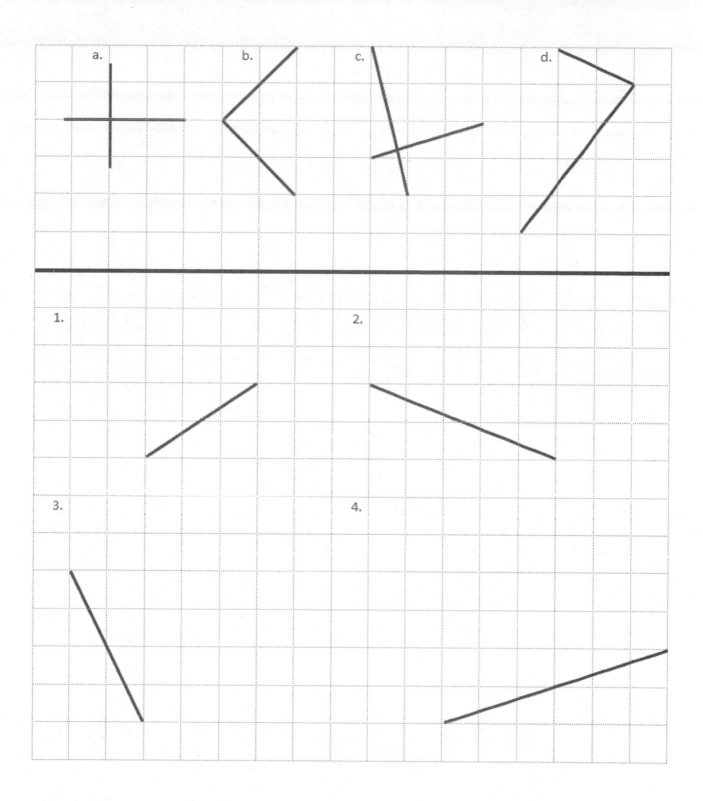

recording sheet

© 2018 Great Minds®. eureka-math.org

a. Complete the table for the rule *y is 1 more than half x*, graph the coordinate pairs, and draw a line to connect them.

b. Give the *y*-coordinate for the point on this line whose *x*-coordinate is $42\frac{1}{4}$.

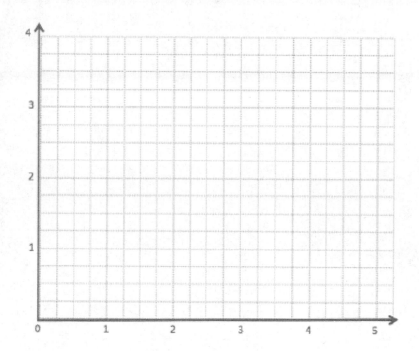

x	y
$\frac{1}{2}$	
$1\frac{1}{2}$	
$2\frac{1}{4}$	
3	

Extension: Give the *x*-coordinate for the point on this line whose *y*-coordinate is $5\frac{1}{2}$.

Read   **Draw**   **Write**

Lesson 16:   Construct perpendicular line segments, and analyze relationships of the coordinate pairs.

133

© 2018 Great Minds®. eureka-math.org

Name _____   Date _____

1. Use the coordinate plane below to complete the following tasks.

    a. Draw $\overline{AB}$.

    b. Plot point C (0, 8).

    c. Draw $\overline{AC}$.

    d. Explain how you know $\angle CAB$ is a
        right angle without measuring it.

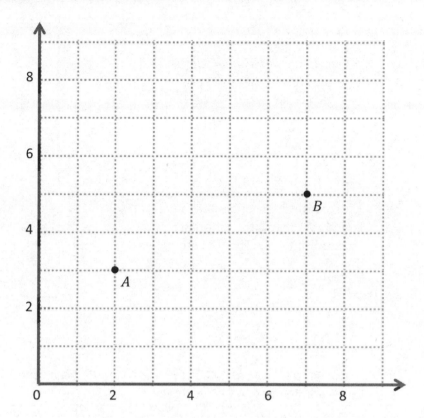

    e. Sean drew the picture below to find a segment perpendicular to $\overline{AB}$. Explain why Sean is correct.

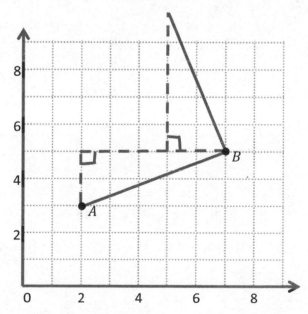

EUREKA
MATH®

Lesson 16:   Construct perpendicular line segments, and analyze relationships of
the coordinate pairs.

135

© 2018 Great Minds®. eureka-math.org

2. Use the coordinate plane below to complete the following tasks.

   a. Draw $\overline{QT}$.

   b. Plot point R $(2, 6\frac{1}{2})$.

   c. Draw $\overline{QR}$.

   d. Explain how you know ∠RQT is a right angle without measuring it.

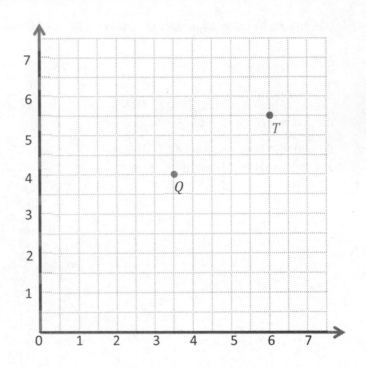

   e. Compare the coordinates of points Q and T. What is the difference of the x-coordinates? The y-coordinates?

   f. Compare the coordinates of points Q and R. What is the difference of the x-coordinates? The y-coordinates?

   g. What is the relationship of the differences you found in parts (e) and (f) to the triangles of which these two segments are a part?

3. $\overleftrightarrow{EF}$ contains the following points. E: (4, 1) F: (8, 7)

   Give the coordinates of a pair of points G and H, such that $\overleftrightarrow{EF} \perp \overline{GH}$.

   G: (_____, _____) H: (_____, _____)

Lesson 16:   Construct perpendicular line segments, and analyze relationships of
               the coordinate pairs.

EUREKA
MATH

© 2018 Great Minds®. eureka-math.org

Name _____   Date _____

Use the coordinate plane below to complete the following tasks.

  a.   Draw $\overline{UV}$.

  b.   Plot point $W(4\frac{1}{2}, 6)$.

  c.   Draw $\overline{VW}$.

  d.   Explain how you know that $\angle UVW$ is a right angle without measuring it.

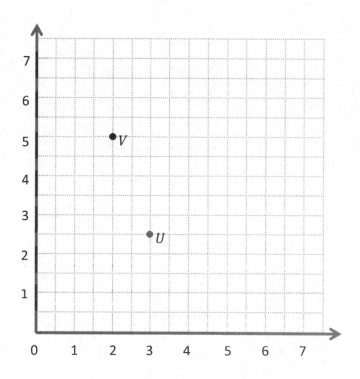

EUREKA
MATH®

Lesson 16:   Construct perpendicular line segments, and analyze relationships of
             the coordinate pairs.

© 2018 Great Minds®. eureka-math.org

137

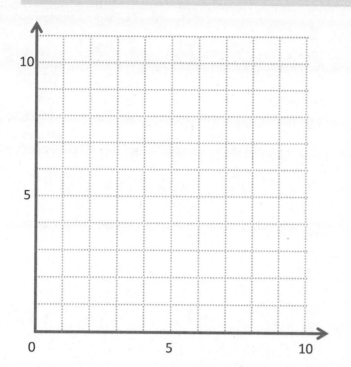

	(x, y)
A	
B	
C	

	(x, y)
D	
E	
F	

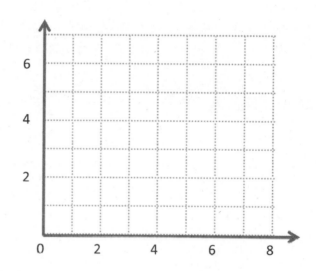

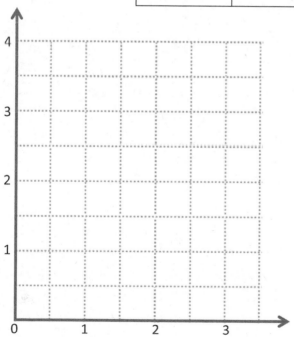

	(x, y)
G	
H	
I	

coordinate plane

EUREKA MATH

Lesson 16:   Construct perpendicular line segments, and analyze relationships of the coordinate pairs.

139

© 2018 Great Minds®. eureka-math.org

Plot (10, 8) and (3, 3) on the coordinate plane, connect the points with a straightedge, and label them as C and D.

a. Draw a segment parallel to $\overline{CD}$.

b. Draw a segment perpendicular to $\overline{CD}$.

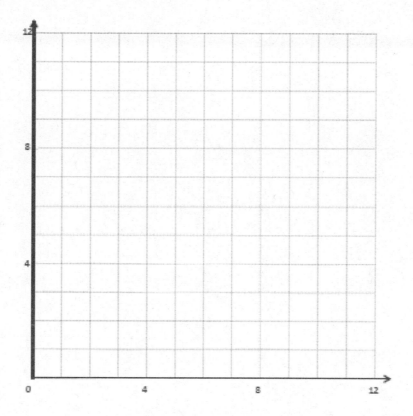

Read   **Draw**   **Write**

Lesson 17:   Draw symmetric figures using distance and angle measure from the
line of symmetry.

141

© 2018 Great Minds®. eureka-math.org

Name _____   Date _____

1. Draw to create a figure that is symmetric about $\overrightarrow{AD}$.

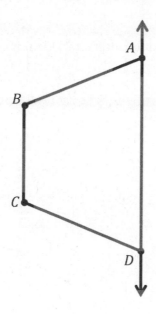

2. Draw precisely to create a figure that is symmetric about $\overleftrightarrow{HI}$.

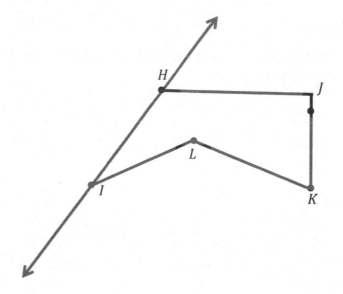

EUREKA MATH

Lesson 17:   Draw symmetric figures using distance and angle measure from the line of symmetry.

143

© 2018 Great Minds®. eureka-math.org

3. Complete the following construction in the space below.

   a. Plot 3 non-collinear points, *D, E*, and *F*.

   b. Draw $\overline{DE}$, $\overrightarrow{DF}$, and $\overline{DF}$.

   c. Plot point *G*, and draw the remaining sides, such that quadrilateral *DEFG* is symmetric about $\overrightarrow{DF}$.

4. Stu says that quadrilateral *HIJK* is symmetric about $\overrightarrow{HJ}$ because *IL* = *LK*. Use your tools to determine Stu's mistake. Explain your thinking.

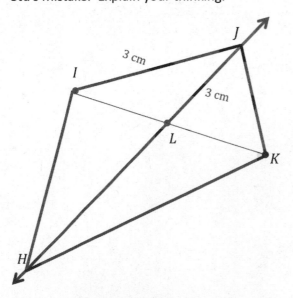

Lesson 17:   Draw symmetric figures using distance and angle measure from the
                         line of symmetry.

                    © 2018 Great Minds®. eureka-math.org

EUREKA
MATH®

Name _____   Date _____

1. Draw 2 points on one side of the line below, and label them T and U.

2. Use your set square and ruler to draw symmetrical points about your line that correspond to T and U, and label them V and W.

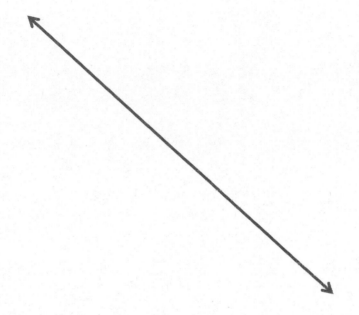

Lesson 17:   Draw symmetric figures using distance and angle measure from the
              line of symmetry.

145

© 2018 Great Minds®. eureka-math.org

Denis buys 8 meters of ribbon. He uses 3.25 meters for a gift. He uses the remaining ribbon equally to tie bows on 5 boxes. How much ribbon did he use on each box?

Read   **Draw**   **Write**

Name _____   Date _____

1. Use the plane to the right to complete the following tasks.

    a. Draw a line *t* whose rule is *y is always 0.7*.

    b. Plot the points from Table A on
        the grid in order. Then, draw line
        segments to connect the points.

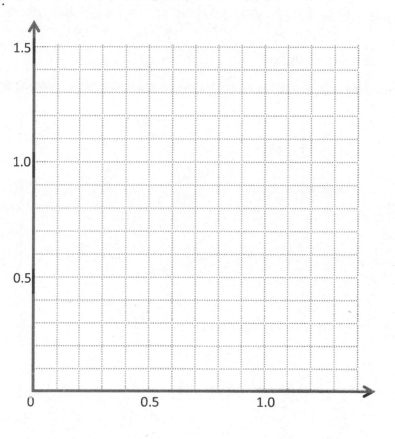

Table A

(x, y)
(0.1, 0.5)
(0.2, 0.3)
(0.3, 0.5)
(0.5, 0.1)
(0.6, 0.2)
(0.8, 0.2)
(0.9, 0.1)
(1.1, 0.5)
(1.2, 0.3)
(1.3, 0.5)

Table B

(x, y)

    c. Complete the drawing to create a figure that is symmetric about line *t*. For each point in Table A,
        record the corresponding point on the other side of the line of symmetry in Table B.

    d. Compare the *y*-coordinates in Table A with those in Table B. What do you notice?

    e. Compare the *x*-coordinates in Table A with those in Table B. What do you notice?

2. This figure has a second line of symmetry. Draw the line on the plane, and write the rule for this line.

3. Use the plane below to complete the following tasks.

   a. Draw a line u whose rule is *y is equal to $x + \frac{1}{4}$*.

   b. Construct a figure with a total of 6 points, all on the same side of the line.

   c. Record the coordinates of each point, in the order in which they were drawn, in Table A.

   d. Swap your paper with a neighbor, and have her complete parts (e–f), below.

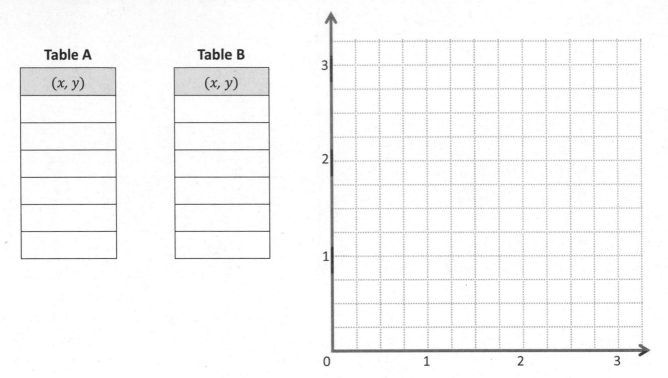

Table A
(x, y)

Table B
(x, y)

   e. Complete the drawing to create a figure that is symmetric about u. For each point in Table A, record the corresponding point on the other side of the line of symmetry in Table B.

   f. Explain how you found the points symmetric to your partner's about u.

EUREKA
MATH

Name _____ Date _____

Kenny plotted the following pairs of points and said they made a symmetric figure about a line with the rule:

y is always 4.

(3, 2) and (3, 6)

(4, 3) and (5, 5)

$(5, \frac{3}{4})$ and $(5, 7\frac{1}{4})$

$(7, 1\frac{1}{2})$ and $(7, 6\frac{1}{2})$

Is his figure symmetrical about the line? How do you know?

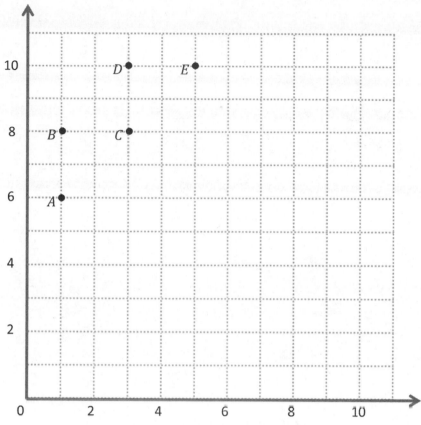

Table A

Point	(x, y)
A	
B	
C	
D	
E	

Table C

(x, y)

Table B

Point	(x, y)
I	
H	
G	
F	

Table D

(x, y)

Table E

Point	(x, y)
A	(1, 1)
B	$(1\frac{1}{2}, 3\frac{1}{2})$
C	(2, 3)
D	$(2\frac{1}{2}, 3\frac{1}{2})$
E	$(2\frac{1}{2}, 2\frac{1}{2})$
F	$(3\frac{1}{2}, 2\frac{1}{2})$
G	(3, 2)
H	$(3\frac{1}{2}, 1\frac{1}{2})$

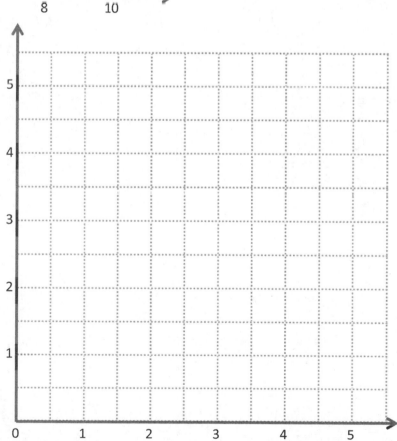

coordinate plane

Three feet are equal to 1 yard. The following table shows the conversion. Use the information to complete the following tasks:

Feet	Yards
3	1
6	2
9	3
12	4

1. Plot each set of coordinates.
2. Use a straightedge to connect each point.
3. Plot one more point on this line, and write its coordinates.

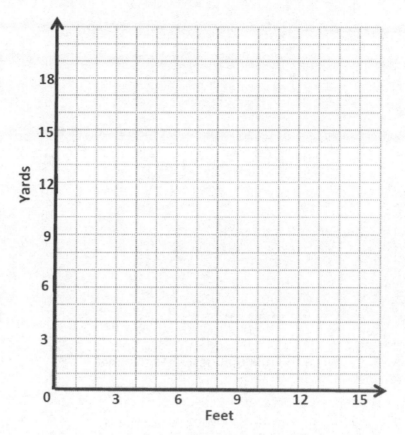

Read   Draw   Write

4. 27 feet can be converted to how many yards?

5. Write the rule that describes the line.

Read   **Draw**   **Write**

Lesson 19:   Plot data on line graphs and analyze trends.

EUREKA
MATH

Name _____ Date _____

1. The line graph below tracks the rain accumulation, measured every half hour, during a rainstorm that began at 2:00 p.m. and ended at 7:00 p.m. Use the information in the graph to answer the questions that follow.

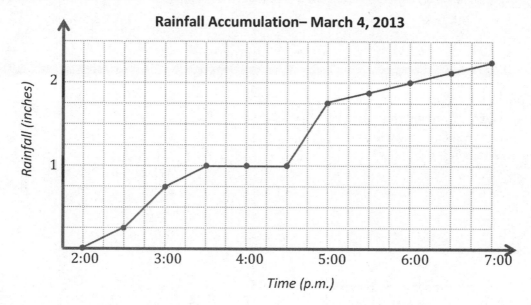

Rainfall Accumulation– March 4, 2013

a. How many inches of rain fell during this five-hour period?

b. During which half-hour period did $\frac{1}{2}$ inch of rain fall? Explain how you know.

c. During which half-hour period did rain fall most rapidly? Explain how you know.

d. Why do you think the line is horizontal between 3:30 p.m. and 4:30 p.m.?

e. For every inch of rain that fell here, a nearby community in the mountains received a foot and a half of snow. How many inches of snow fell in the mountain community between 5:00 p.m. and 7:00 p.m.?

2. Mr. Boyd checks the gauge on his home's fuel tank on the first day of every month. The line graph to the right was created using the data he collected.

a. According to the graph, during which month(s) does the amount of fuel decrease most rapidly?

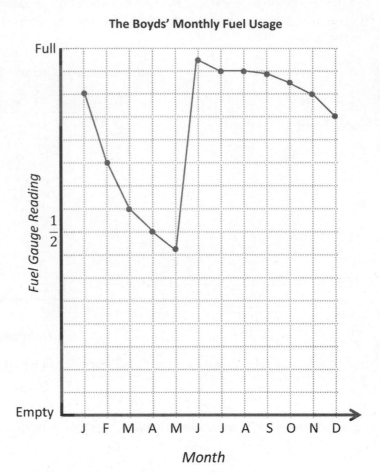

The Boyds' Monthly Fuel Usage

b. The Boyds took a month-long vacation. During which month did this most likely occur? Explain how you know using the data in the graph.

c. Mr. Boyd's fuel company filled his tank once this year. During which month did this most likely occur? Explain how you know.

d. The Boyd family's fuel tank holds 284 gallons of fuel when full. How many gallons of fuel did the Boyds use in February?

e. Mr. Boyd pays $3.54 per gallon of fuel. What is the cost of the fuel used in February and March?

Lesson 19: Plot data on line graphs and analyze trends.

EUREKA
MATH

Name _____   Date _____

The line graph below tracks the water level of Plainsview Creek, measured each Sunday, for 8 weeks. Use the information in the graph to answer the questions that follow.

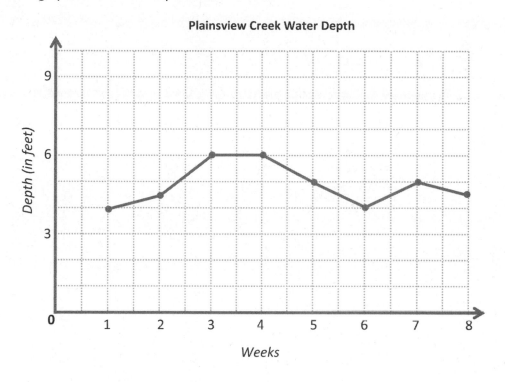

a. About how many feet deep was the creek in Week 1? _____

b. According to the graph, which week had the greatest change in water depth? _____

c. It rained hard throughout the sixth week. During what other weeks might it have rained? Explain why you think so.

d. What might have been another cause leading to an increase in the depth of the creek?

Name _____   Date _____

1. The line graph below tracks the total tomato production for one tomato plant. The total tomato production is plotted at the end of each of 8 weeks. Use the information in the graph to answer the questions that follow.

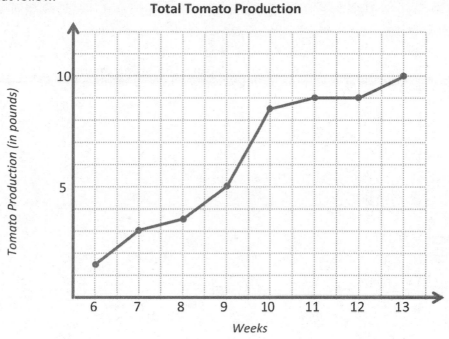

a. How many pounds of tomatoes did this plant produce at the end of 13 weeks?

b. How many pounds of tomatoes did this plant produce from Week 7 to Week 11? Explain how you know.

c. Which one-week period showed the greatest change in tomato production? The least? Explain how you know.

d. During Weeks 6–8, Jason fed the tomato plant just water. During Weeks 8–10, he used a mixture of water and Fertilizer A, and in Weeks 10–13, he used water and Fertilizer B on the tomato plant. Compare the tomato production for these periods of time.

2. Use the story context below to sketch a line graph. Then, answer the questions that follow.

The number of fifth-grade students attending Magnolia School has changed over time. The school opened in 2006 with 156 students in the fifth grade. The student population grew the same amount each year before reaching its largest class of 210 students in 2008. The following year, Magnolia lost one-seventh of its fifth graders. In 2010, the enrollment dropped to 154 students and remained constant in 2011. For the next two years, the enrollment grew by 7 students each year.

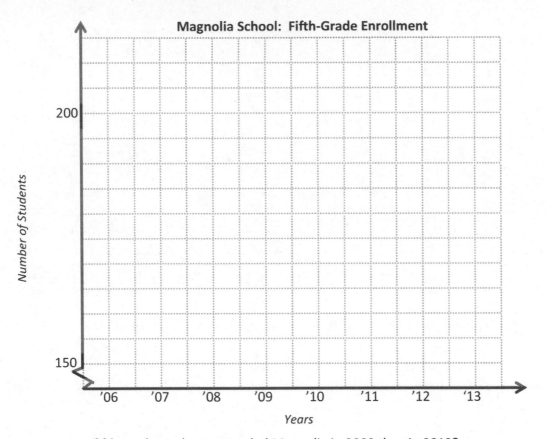

a. How many more fifth-grade students attended Magnolia in 2009 than in 2013?

b. Between which two consecutive years was there the greatest change in student population?

c. If the fifth-grade population continues to grow in the same pattern as in 2012 and 2013, in what year will the number of students match 2008's enrollment?

Lesson 20:   Use coordinate systems to solve real-world problems.   EUREKA
                                                                                  MATH

Name _____ Date _____

Use the following information to complete the line graph below. Then, answer the questions that follow.

Harry runs a hot dog stand at the county fair. When he arrived on Wednesday, he had 38 dozen hot dogs for his stand. The graph shows the number of hot dogs (in dozens) that remained unsold at the end of each day of sales.

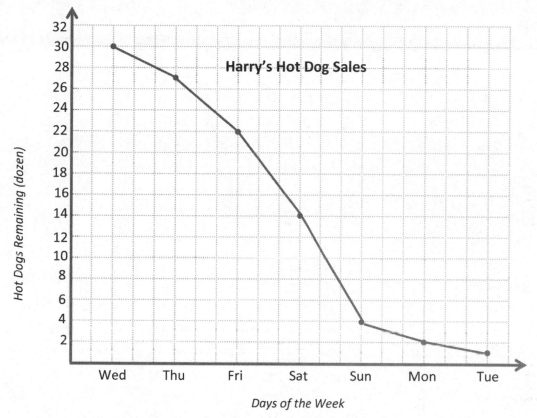

a.   How many dozen hot dogs did Harry sell on Wednesday? How do you know?

b.   Between which two-day period did the number of hot dogs sold change the most? Explain how you determined your answer.

c.   During which three days did Harry sell the most hot dogs?

d.   How many dozen hot dogs were sold on these three days?

Student _____ Team _____ Date _____ Problem 1

Pierre's Paper

Pierre folded a square piece of paper vertically to make two rectangles. Each rectangle had a perimeter of 39 inches. How long is each side of the original square? What is the area of the original square? What is the area of one of the rectangles?

Student _____ Team _____ Date _____ Problem 2

Shopping with Elise

Elise saved $184. She bought a scarf, a necklace, and a notebook. After her purchases, she still had $39.50. The scarf cost three-fifths the cost of the necklace, and the notebook was one-sixth as much as the scarf. What was the cost of each item? How much more did the necklace cost than the notebook?

Lessons 21–23: Make sense of complex, multi-step problems, and persevere in solving them. Share and critique peer solutions.

165

© 2018 Great Minds®. eureka-math.org

Student _____ Team _____ Date _____ Problem 3

The Hewitt's Carpet

The Hewitt family is buying carpet for two rooms. The dining room is a square that measures 12 feet on each side. The den is 9 yards by 5 yards. Mrs. Hewitt has budgeted $2,650 for carpeting both rooms. The green carpet she is considering costs $42.75 per square yard, and the brown carpet's price is $4.95 per square foot. What are the ways she can carpet the rooms and stay within her budget?

Student _____ Team _____ Date _____ Problem 4

AAA Taxi

AAA Taxi charges $1.75 for the first mile and $1.05 for each additional mile. How far could Mrs. Leslie travel for $20 if she tips the cab driver $2.50?

     Lessons 21–23:   Make sense of complex, multi-step problems, and persevere in solving them. Share and critique peer solutions.

© 2018 Great Minds®. eureka-math.org

Student _____ Team _____ Date _____ Problem 5

Pumpkins and Squash

Three pumpkins and two squash weigh 27.5 pounds. Four pumpkins and three squash weigh 37.5 pounds. Each pumpkin weighs the same as the other pumpkins, and each squash weighs the same as the other squash. How much does each pumpkin weigh? How much does each squash weigh?

Student _____ Team _____ Date _____ Problem 6

Toy Cars and Trucks

Henry had 20 convertibles and 5 trucks in his miniature car collection. After Henry's aunt bought him some more miniature trucks, Henry found that one-fifth of his collection consisted of convertibles. How many trucks did his aunt buy?

Lessons 21–23:   Make sense of complex, multi-step problems, and persevere in solving them. Share and critique peer solutions.   **167**

© 2018 Great Minds®. eureka-math.org

Student _____ Team _____ Date _____ Problem 7

Pairs of Scouts

Some girls in a Girl Scout troop are pairing up with some boys in a Boy Scout troop to practice square dancing. Two-thirds of the girls are paired with three-fifths of the boys. What fraction of the scouts are square dancing?

(Each pair is one Girl Scout and one Boy Scout. The pairs are only from these two troops.)

Student _____ Team _____ Date _____ Problem 8

Sandra's Measuring Cups

Sandra is making cookies that require $5\frac{1}{2}$ cups of oatmeal. She has only two measuring cups: a one-half cup and a three-fourths cup. What is the smallest number of scoops that she could make in order to get $5\frac{1}{2}$ cups?

Lessons 21–23:   Make sense of complex, multi-step problems, and persevere in solving
them. Share and critique peer solutions.

Student _____ Team _____ Date _____ Problem 9

Blue Squares

The dimensions of each successive blue square pictured to the right are half that of the previous blue square. The lower left blue square measures 6 inches by 6 inches.

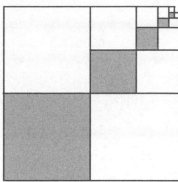

    a. Find the area of the shaded part.

    b. Find the total area of the shaded and unshaded parts.

    c. What fraction of the figure is shaded?

Lessons 21–23: Make sense of complex, multi-step problems, and persevere in solving
them. Share and critique peer solutions.

169

© 2018 Great Minds®. eureka-math.org

The market sells watermelons for $0.39 per pound and apples for $0.43 per pound. Write an expression that shows how much Carmen spends for a watermelon that weighs 11.5 pounds and a bag of apples that weighs 3.2 pounds.

_____

_____

_____

_____

Read   **Draw**   **Write**

Name _____ Date _____

1. For each written phrase, write a numerical expression, and then evaluate your expression.

   a. Three fifths of the sum of thirteen and six

      Numerical expression:

      Solution:

   b. Subtract four thirds from one seventh of sixty-three.

      Numerical expression:

      Solution:

   c. Six copies of the sum of nine fifths and three

      Numerical expression:

      Solution:

   d. Three fourths of the product of four fifths and fifteen

      Numerical expression:

      Solution:

2. Write at least 2 numerical expressions for each phrase below. Then, solve.

    a. Two thirds of eight

    b. One sixth of the product of four and nine

3. Use <, >, or = to make true number sentences without calculating. Explain your thinking.

    a. $217 \times (42 + \frac{48}{5})$ $(217 \times 42) + \frac{48}{5}$

    b. $(687 \times \frac{3}{16}) \times \frac{7}{12}$ $(687 \times \frac{3}{16}) \times \frac{3}{12}$

    c. $5 \times 3.76 + 5 \times 2.68$ ◯ 5×6.99

     Lesson 26:   Solidify writing and interpreting numerical expressions.

EUREKA
MATH®

six sevenths of nine	two thirds the sum of twenty-three and fifty-seven	forty-three less than three fifths of the product of ten and twenty	five sixths the difference of three hundred twenty-nine and two hundred eighty-one
three times as much as the sum of three fourths and two thirds	the difference between thirty thirties and twenty-eight thirties	twenty-seven more than half the sum of four and one eighth and six and two thirds	the sum of eighty-eight and fifty-six divided by twelve
the product of nine and eight divided by four	one sixth the product of twelve and four	six copies of the sum of six twelfths and three fourths	double three fourths of eighteen

expression cards

$96 \times (63 + \frac{17}{12})$ $(96 \times 63) + \frac{17}{12}$

$(437 \times \frac{9}{15}) \times \frac{6}{8}$ $(437 \times \frac{9}{15}) \times \frac{7}{8}$

$4 \times 8.35 + 4 \times 6.21$ 4×15.87

$\frac{6}{7} \times (3{,}065 + 4{,}562)$ $(3{,}065 + 4{,}562) + \frac{6}{7}$

$(8.96 \times 3) + (5.07 \times 8)$ $(8.96 + 3) \times (5.07 + 8)$

$(297 \times \frac{16}{15}) + \frac{8}{3}$ $(297 \times \frac{13}{15}) + \frac{8}{3}$

$\frac{12}{7} \times (\frac{5}{4} + \frac{5}{9})$ $\frac{12}{7} \times \frac{5}{4} + \frac{12}{7} \times \frac{5}{9}$

comparing expressions game board

Name _____ Date _____

1. Use the RDW process to solve the word problems below.

   a. Julia completes her homework in an hour. She spends $\frac{7}{12}$ of the time doing her math homework and $\frac{1}{6}$ of the time practicing her spelling words. The rest of the time she spends reading. How many minutes does Julia spend reading?

   b. Fred has 36 marbles. Elise has $\frac{8}{9}$ as many marbles as Fred. Annika has $\frac{3}{4}$ as many marbles as Elise. How many marbles does Annika have?

2. Write and solve a word problem that might be solved using the expressions in the chart below.

Expression	Word Problem	Solution
$\dfrac{2}{3} \times 18$		
$(26 + 34) \times \dfrac{5}{6}$		
$7 - \left(\dfrac{5}{12} + \dfrac{1}{2}\right)$		

Lesson 27:   Solidify writing and interpreting numerical expressions.

EUREKA MATH

Name _____ Date _____

1. Answer the following questions about fluency.

   a. What does being fluent with a math skill mean to you?

   b. Why is fluency with certain math skills important?

   c. With which math skills do you think you should be fluent?

   d. With which math skills do you feel most fluent? Least fluent?

   e. How can you continue to improve your fluency?

2. Use the chart below to list skills from today's activities with which you are fluent.

Fluent Skills

3. Use the chart below to list skills we practiced today with which you are less fluent.

Skills to Practice More

Write Fractions as Mixed Numbers

Materials: (S) Personal white board

T: (Write $\frac{13}{2}$ = _____ ÷ _____ = _____ .) Write the fraction as a division problem and mixed number.

S: (Write $\frac{13}{2}$ = 13 ÷ 2 = $6\frac{1}{2}$.)

More practice!

$\frac{11}{2}, \frac{17}{2}, \frac{44}{2}, \frac{31}{10}, \frac{23}{10}, \frac{47}{10}, \frac{89}{10}, \frac{8}{3}, \frac{13}{3}, \frac{26}{3}, \frac{9}{4}, \frac{13}{4}, \frac{15}{4}$, and $\frac{35}{4}$.

Fraction of a Set

Materials: (S) Personal white board

T: (Write $\frac{1}{2} \times 10$.) Draw a tape diagram to model the whole number.

S: (Draw a tape diagram, and label it 10.)

T: Draw a line to split the tape diagram in half.

S: (Draw a line.)

T: What is the value of each part of your tape diagram?

S: 5.

T: So, what is $\frac{1}{2}$ of 10?

S: 5.

More practice!

$8 \times \frac{1}{2}$, $8 \times \frac{1}{4}$, $6 \times \frac{1}{3}$, $30 \times \frac{1}{6}$, $42 \times \frac{1}{7}$, $42 \times \frac{1}{6}$, $48 \times \frac{1}{8}$,

$54 \times \frac{1}{9}$, and $54 \times \frac{1}{6}$.

Convert to Hundredths

Materials: (S) Personal white board

T: (Write $\frac{3}{4} = \frac{}{100}$.) 4 times what factor equals 100?

S: 25.

T: Write the equivalent fraction.

S: (Write $\frac{3}{4} = \frac{75}{100}$.)

More practice!

$\frac{3}{4} = \frac{}{100}$, $\frac{1}{50} = \frac{}{100}$, $\frac{3}{50} = \frac{}{100}$, $\frac{1}{20} = \frac{}{100}$, $\frac{3}{20} = \frac{}{100}$,

$\frac{1}{25} = \frac{}{100}$, and $\frac{2}{25} = \frac{}{100}$.

Multiply a Fraction and a Whole Number

Materials: (S) Personal white board

T: (Write $\frac{8}{4}$.) Write the corresponding division sentence.

S: (Write 8 ÷ 4 = 2.)

T: (Write $\frac{1}{4} \times 8$.) Write the complete multiplication sentence.

S: (Write $\frac{1}{4} \times 8 = 2$.)

More practice!

$\frac{18}{6}, \frac{15}{3}, \frac{18}{3}, \frac{27}{9}, \frac{54}{6}, \frac{51}{3}$, and $\frac{63}{7}$.

fluency activities

Multiply Mentally

Materials: (S) Personal white board

T: (Write 9 × 10.) On your personal white board, write the complete multiplication sentence.
S: (Write 9 × 10 = 90.)
T: (Write 9 × 9 = 90 − _____ below 9 × 10 = 90.) Write the number sentence, filling in the blank.
S: (Write 9 × 9 = 90 − 9.)
T: 9 × 9 is…?
S: 81.

More practice!

9 × 99, 15 × 9, and 29 × 99.

One Unit More

Materials: (S) Personal white board

T: (Write 5 tenths.) On your personal white board, write the decimal that's one-tenth more than 5 tenths.
S: (Write 0.6.)

More practice!

5 hundredths, 5 thousandths, 8 hundredths, and 2 thousandths. Specify the unit of increase.

T: (Write 0.052.) Write one more thousandth.
S: (Write 0.053.)

More practice!

1 tenth more than 35 hundredths,
1 thousandth more than 35 hundredths, and
1 hundredth more than 438 thousandths.

Find the Product

Materials: (S) Personal white board

T: (Write 4 × 3.) Complete the multiplication sentence giving the second factor in unit form.
S: (Write 4 × 3 ones = 12 ones.)
T: (Write 4 × 0.2.) Complete the multiplication sentence giving the second factor in unit form.
S: (Write 4 × 2 tenths = 8 tenths.)
T: (Write 4 × 3.2.) Complete the multiplication sentence giving the second factor in unit form.
S: (Write 4 × 3 ones 2 tenths = 12 ones 8 tenths.)
T: Write the complete multiplication sentence.
S: (Write 4 × 3.2 = 12.8.)

More practice!

4 × 3.21, 9 × 2, 9 × 0.1, 9 × 0.03, 9 × 2.13, 4.012 × 4, and 5 × 3.2375.

Add and Subtract Decimals

Materials: (S) Personal white board

T: (Write 7 ones + 258 thousandths + 1 hundredth = _____.) Write the addition sentence in decimal form.
S: (Write 7 + 0.258 + 0.01 = 7.268.)

More practice!

7 ones + 258 thousandths + 3 hundredths,
6 ones + 453 thousandths + 4 hundredths,
2 ones + 37 thousandths + 5 tenths, and
6 ones + 35 hundredths + 7 thousandths.

T: (Write 4 ones + 8 hundredths − 2 ones = _____ ones _____ hundredths.) Write the subtraction sentence in decimal form.
S: (Write 4.08 − 2 = 2.08.)

More practice!

9 tenths + 7 thousandths − 4 thousandths,
4 ones + 582 thousandths − 3 hundredths,
9 ones + 708 thousandths − 4 tenths, and
4 ones + 73 thousandths − 4 hundredths.

fluency activities

Decompose Decimals

Materials: (S) Personal white board

T: (Project 7.463.)
     Say the number.

S: 7 and 463
     thousandths.

T: Represent this
     number in a
     two-part number
     bond with ones
     as one part and
     thousandths as
     the other part.

S: (Draw.)

T: Represent it again
     with tenths and
     thousandths.

S: (Draw.)

T: Represent it again with
     hundredths and thousandths.

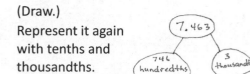

More practice!

8.972 and 6.849.

Find the Volume

Materials: (S) Personal white board

T: On your personal white board, write
     the formula for finding the volume
     of a rectangular prism.

S: (Write $V = l \times w \times h$.)

T: (Draw and label a rectangular prism with
     a length of 5 cm, width of 6 cm, and height
     of 2 cm.) Write a multiplication sentence
     to find the volume of this rectangular prism.

S: (Beneath $V = l \times w \times h$, write V = 5 cm ×
     6 cm × 2 cm. Beneath it, write V = 60 cm³.)

More practice!

l = 7 ft, w = 9 ft, h = 3 ft;

l = 6 in, w = 6 in, h = 5 in; and

l = 4 cm, w = 8 cm, h = 2 cm.

Make a Like Unit

Materials: (S) Personal white board

T: I will say two unit fractions. You make
     the like unit, and write it on your
     personal white board. Show your
     board at the signal.

T: $\frac{1}{3}$ and $\frac{1}{2}$. (Pause. Signal.)

S: (Write and show sixths.)

More practice!

$\frac{1}{4}$ and $\frac{1}{3}$, $\frac{1}{2}$ and $\frac{1}{4}$, $\frac{1}{6}$ and $\frac{1}{2}$, $\frac{1}{3}$ and $\frac{1}{12}$, $\frac{1}{6}$ and
$\frac{1}{8}$, and $\frac{1}{3}$ and $\frac{1}{9}$.

Unit Conversions

Materials: (S) Personal white board

T: (Write 12 in = _____ ft.) On your personal
     white board, write 12 inches is the same
     as how many feet?

S: (Write 1 foot.)

More practice!

24 in, 36 in, 54 in, and 76 in.

T: (Write 1 ft = _____ in.) Write 1 foot is the same
     as how many inches?

S: (Write 12 inches.)

More practice!

2 ft, 2.5 ft, 3 ft, 3.5 ft, 4 ft, 4.5 ft, 9 ft, and 9.5 ft.

fluency activities

Lesson 28:   Solidify fluency with Grade 5 skills.

Compare Decimal Fractions

Materials: (S) Personal white board

T: (Write 13.78 ___ 13.86.) On your personal white board, compare the numbers using the greater than, less than, or equal sign.

S: (Write 13.78 < 13.86.)

More practice!

0.78____$\frac{78}{100}$, 439.3 ___ 4.39, 5.08 ___ fifty-eight tenths, and thirty-five and 9 thousandths ___ 4 tens.

Round to the Nearest One

Materials: (S) Personal white board

T: (Write 3 ones 2 tenths.) Write 3 ones and 2 tenths as a decimal.

S: (Write 3.2.)

T: (Write $3.2 \approx$ __.) Round 3 and 2 tenths to the nearest whole number.

S: (Write $3.2 \approx 3$.)

More practice!

3.7, 13.7, 5.4, 25.4, 1.5, 21.5, 6.48, 3.62, and 36.52.

Multiplying Fractions

Materials: (S) Personal white board

T: (Write $\frac{1}{2} \times \frac{1}{3} =$ ___.) Write the complete multiplication sentence.

S: (Write $\frac{1}{2} \times \frac{1}{3} = \frac{1}{6}$.)

T: (Write $\frac{1}{2} \times \frac{3}{4} =$ ___.) Write the complete multiplication sentence.

S: (Write $\frac{1}{2} \times \frac{3}{4} = \frac{3}{8}$.)

T: (Write $\frac{2}{5} \times \frac{2}{3} =$ ___.) Write the complete multiplication sentence.

S: (Write $\frac{2}{5} \times \frac{2}{3} = \frac{4}{15}$.)

More practice!

$\frac{1}{2} \times \frac{1}{5}$, $\frac{1}{2} \times \frac{3}{5}$, $\frac{3}{4} \times \frac{3}{5}$, $\frac{4}{5} \times \frac{2}{3}$, and $\frac{3}{4} \times \frac{5}{6}$.

Divide Numbers by Unit Fractions

Materials: (S) Personal white board

T: (Write $1 \div \frac{1}{2}$.) How many halves are in 1?

S: 2.

T: (Write $1 \div \frac{1}{2} = 2$. Beneath it, write $2 \div \frac{1}{2}$.) How many halves are in 2?

S: 4.

T: (Write $2 \div \frac{1}{2} = 4$. Beneath it, write $3 \div \frac{1}{2}$.) How many halves are in 3?

S: 6.

T: (Write $3 \div \frac{1}{2} = 6$. Beneath it, write $7 \div \frac{1}{2}$.) Write the complete division sentence.

S: (Write $7 \div \frac{1}{2} = 14$.)

More practice!

$1 \div \frac{1}{3}$, $2 \div \frac{1}{5}$, $9 \div \frac{1}{4}$, and $3 \div \frac{1}{8}$.

fluency activities

A quadrilateral with two pairs of equal sides that are also adjacent.	An angle that turns through $\frac{1}{360}$ of a circle.	A quadrilateral with at least one pair of parallel lines.	A closed figure made up of line segments.
Measurement of space or capacity.	A quadrilateral with opposite sides that are parallel.	An angle measuring 90 degrees.	The union of two different rays sharing a common vertex.
The number of square units that cover a two-dimensional shape.	Two lines in a plane that do not intersect.	The number of adjacent layers of the base that form a rectangular prism.	A three-dimensional figure with six square sides.
A quadrilateral with four 90-degree angles.	A polygon with 4 sides and 4 angles.	A parallelogram with all equal sides.	Cubes of the same size used for measuring.
Two intersecting lines that form 90-degree angles.	A three-dimensional figure with six rectangular sides.	A three-dimensional figure.	Any flat surface of a 3-D figure.
A line that cuts a line segment into two equal parts at 90 degrees.	Squares of the same size, used for measuring.	A rectangular prism with only 90-degree angles.	One face of a 3-D solid, often thought of as the surface upon which the solid rests.

geometry definitions

Base	Volume of a Solid	Cubic Units	Kite
Height	One-Degree Angle	Face	Trapezoid
Right Rectangular Prism	Perpendicular Bisector	Cube	Area
Perpendicular Lines	Rhombus	Parallel Lines	Angle
Polygon	Rectangular Prism	Parallelogram	Rectangle
Right Angle	Quadrilateral	Solid Figure	Square Units

geometry terms

Attribute Buzz:

Number of players: 2

Description: Players place geometry terms cards facedown in a pile and, as they select cards, name the attributes of each figure within 1 minute.

- Player A flips the first card and says as many attributes as possible within 30 seconds.

- Player B says, "Buzz," when or if Player A states an incorrect attribute or time is up.

- Player B explains why the attribute is incorrect (if applicable) and can then start listing attributes about the figure for 30 seconds.

- Players score a point for each correct attribute.

- Play continues until students have exhausted the figure's attributes. A new card is selected, and play continues. The player with the most points at the end of the game wins.

Concentration:

Number of players: 2–6

Description: Players persevere to match term cards with their definition and description cards.

- Create two identical arrays side by side: one of term cards and one of definition and description cards.

- Players take turns flipping over pairs of cards to find a match. A match is a vocabulary term and its definition or description card. Cards keep their precise location in the array if not matched. Remaining cards are not reconfigured into a new array.

- After all cards are matched, the player with the most pairs is the winner.

Three Questions to Guess My Term!

Number of players: 2–4

Description: A player selects and secretly views a term card. Other players take turns asking yes or no questions about the term.

- Players can keep track of what they know about the term on paper.

- Only yes or no questions are allowed. ("What kind of angles do you have?" is not allowed.)

- A final guess must be made after 3 questions but may be made sooner. Once a player says, "This is my guess," no more questions may be asked by that player.

- If the term is guessed correctly after 1 or 2 questions, 2 points are earned. If all 3 questions are used, only 1 point is earned.

- If no player guesses correctly, the card holder receives the point.

- The game continues as the player to the card holder's left selects a new card and questioning begins again.

- The game ends when a player reaches a predetermined score.

Bingo:

Number of players: at least 4—whole class

Description: Players match definitions to terms to be the first to fill a row, column, or diagonal.

- Players write a geometry term in each box of the math bingo card. Each term should be used only once. The box that says *Math Bingo!* is a free space.

- Players place the filled-in math bingo template in their personal white boards.

- One person is the caller and reads the definition from a geometry definition card.

- Players cross off or cover the term that matches the definition.

- Bingo!" is called when 5 vocabulary terms in a row are crossed off diagonally, vertically, or horizontally. The free space counts as 1 box toward the needed 5 vocabulary terms.

- The first player to have 5 in a row reads each crossed-off word, states the definition, and gives a description or an example of each word. If all words are reasonably explained as determined by the caller, the player is declared the winner.

game directions

		Math BINGO		

		Math BINGO		

bingo card

Step 1   Draw $\overline{AB}$ 3 inches long centered near the bottom of a blank piece of paper.

Step 2   Draw $\overline{AC}$ 3 inches long, such that ∠ BAC measures 108°.

Step 3   Draw $\overline{CD}$ 3 inches long, such that ∠ ACD measures 108°.

Step 4   Draw $\overline{DE}$ 3 inches long, such that ∠ CDE measures 108°.

Step 5   Draw $\overline{EB}$.

Step 6   Measure $\overline{EB}$.

_____

_____

_____

_____

Read   **Draw**   **Write**

Name _____ Date _____

Write the Fibonacci sequence. Analyze which numbers are even. Is there a pattern to the even numbers? Why? Think about the spiral of squares that you made yesterday.

_____

_____

_____

_____

Read   **Draw**   **Write**

Name _____   Date _____

1. Ashley decides to save money, but she wants to build it up over a year. She starts with $1.00 and adds 1 more dollar each week. Complete the table to show how much she will have saved after a year.

Week	Add	Total	Week	Add	Total
1	$1.00	$1.00	27		
2	$2.00	$3.00	28		
3	$3.00	$6.00	29		
4	$4.00	$10.00	30		
5			31		
6			32		
7			33		
8			34		
9			35		
10			36		
11			37		
12			38		
13			39		
14			40		
15			41		
16			42		
17			43		
18			44		
19			45		
20			46		
21			47		
22			48		
23			49		
24			50		
25			51		
26			52		

2. Carly wants to save money, too, but she has to start with the smaller denomination of quarters. Complete the second chart to show how much she will have saved by the end of the year if she adds a quarter more each week. Try it yourself, if you can and want to!

Week	Add	Total	Week	Add	Total
1	$2.25	$0.25	27		
2	$0.50	$0.75	28		
3	$0.75	$1.50	29		
4	$1.00	$2.50	30		
5			31		
6			32		
7			33		
8			34		
9			35		
10			36		
11			37		
12			38		
13			39		
14			40		
15			41		
16			42		
17			43		
18			44		
19			45		
20			46		
21			47		
22			48		
23			49		
24			50		
25			51		
26			52		

Lesson 32: Explore patterns in saving money.

EUREKA MATH®

3. David decides he wants to save even more money than Ashley did. He does so by adding the next Fibonacci number instead of adding $1.00 each week. Use your calculator to fill in the chart and find out how much money he will have saved by the end of the year. Is this realistic for most people? Explain your answer.

Week	Add	Total	Week	Add	Total
1	$1	$1	27		
2	$1	$2	28		
3	$2	$4	29		
4	$3	$7	30		
5	$5	$12	31		
6	$8	$20	32		
7			33		
8			34		
9			35		
10			36		
11			37		
12			38		
13			39		
14			40		
15			41		
16			42		
17			43		
18			44		
19			45		
20			46		
21			47		
22			48		
23			49		
24			50		
25			51		
26			52		

Name _____ Date _____

Record the dimensions of your boxes and lid below. Explain your reasoning for the dimensions you chose for Box 2 and the lid.

BOX 1 (Can hold Box 2 inside.)

The dimensions of Box 1 are _____ × _____ × _____.

Its volume is _____.

BOX 2 (Fits inside of Box 1.)

The dimensions of Box 2 are _____ × _____ × _____.

Reasoning:

LID (Fits snugly over Box 1 to protect the contents.)

The dimensions of the lid are _____ × _____ × _____.

Reasoning:

1. What steps did you take to determine the dimensions of the lid?

2. Find the volume of Box 2. Then, find the difference in the volumes of Boxes 1 and 2.

3. Imagine Box 3 is created such that each dimension is 1 cm less than that of Box 2. What would the volume of Box 3 be?

Lesson 33:   Design and construct boxes to house materials for summer use.

EUREKA
MATH

Steven is a _____ who had $280. He spent $\frac{1}{4}$ of his money on a _____ and $\frac{5}{6}$ of the remainder on a _____. How much money did he spend altogether?

_____

_____

_____

_____

Read   **Draw**   **Write**

Name _____ Date _____

I reviewed _____'s work.

Use the chart below to evaluate your friend's two boxes and lid. Measure and record the dimensions, and calculate the box volumes. Then, assess suitability, and suggest improvements in the adjacent columns.

Dimensions and Volume	Is the Box or Lid Suitable? Explain.	Suggestions for Improvement
BOX 1 dimensions: Total volume:		
BOX 2 dimensions: Total volume:		
LID dimensions: 		

Lesson 34:   Design and construct boxes to house materials for summer use.

213

© 2018 Great Minds®. eureka-math.org

Credits

Great Minds® has made every effort to obtain permission for the reprinting of all copyrighted material. If any owner of copyrighted material is not acknowledged herein, please contact Great Minds for proper acknowledgment in all future editions and reprints of this module.